The Search for Our Cosmic Ancestry

The Search for Our Cosmic Ancestry

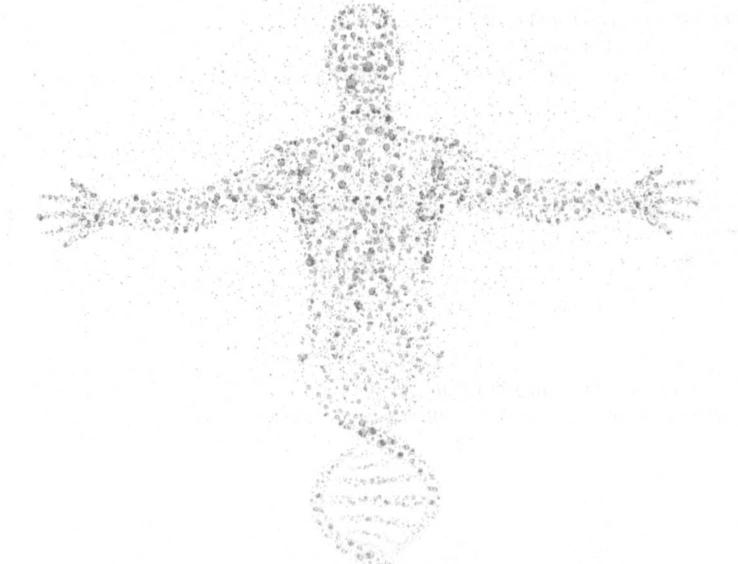

Chandra Wickramasinghe

Buckingham Centre for Astrobiology, UK

World Scientific

NEW JERSEY • LONDON • SINGAPORE • BEIJING • SHANGHAI • HONG KONG • TAIPEI • CHENNAI

Published by

World Scientific Publishing Co. Pte. Ltd.
5 Toh Tuck Link, Singapore 596224
USA office: 27 Warren Street, Suite 401-402, Hackensack, NJ 07601
UK office: 57 Shelton Street, Covent Garden, London WC2H 9HE

Library of Congress Cataloging-in-Publication Data
Wickramasinghe, Chandra, 1939– author.
The search for our cosmic ancestry / Nalin Chandra Wickramasinghe, Buckingham Centre for Astrobiology, UK.
pages cm
Includes bibliographical references and index.
ISBN 978-9814616966 (hardcover : alk. paper) -- ISBN 978-9814616973 (softcover : alk. paper)
1. Exobiology. 2. Extraterrestrial beings. 3. Life on other planets. I. Title.
QH326.W534 2014
576.8'39--dc23

2014018316

British Library Cataloguing-in-Publication Data
A catalogue record for this book is available from the British Library.

In-house Editor: Ng Kah Fee

Printed in Singapore

Dedication

In memory of Fred Hoyle who changed the way we perceive our connection with the external Universe more profoundly than anyone in living memory.

Preface

Since Fred Hoyle's passing in 2001, evidence supporting the theory of cometary panspermia has grown to the point of becoming almost compelling. The most direct supportive evidence has come from the discovery of biological structures within cometary fragments and meteoroids arriving at the Earth. Astronomical evidence for organic molecules linked to biology has now been extended to include galaxies located at distances exceeding 11 billion light years, thus indicating that life is distributed throughout a very large fraction of the observable Universe. Evidence from DNA sequencing studies has revealed that our DNA contains sequences belonging to viruses that were evidently the result of past infections that inevitably contributed to later evolution of our species. A closed-box neo-Darwinian evolutionary paradigm is rapidly giving way to the type of model proposed by Hoyle and the present author in their 1980 book *Evolution from Space*. Terrestrial evolution is the expression on the Earth of a long history of developments that has taken place across vast cosmological distances, and over aeons of time.

Chandra Wickramasinghe

Cardiff, March 2014

Contents

List of Tables

List of Figures

Prologue

The problems facing humanity in 2014 are concerned with survival itself. The world population is rising at a rate that would eventually compromise our capacity to sustain ourselves with the food, energy and other resources that we have been accustomed to enjoy. Climate change and the quest for sources of renewable energy pose a continuing challenge that is becoming ever more serious. Not unconnected with such basic problems is a growing tendency of humans to revert to a level of primitive hostility as evidenced in the ongoing confrontations between rival creeds and nations, each seeking supremacy and constantly having to resist deployment of the most lethal weapons of war. Is this a reversion to our evolutionary past when such strife for survival was an imperative need? Or is it an instability resulting from uncontrolled technological progress? To resolve such questions it is important that we turn to science to acquire a true understanding of the problems we face.

There are several kinds of science to consider in this context. Firstly there is rigorous science based directly upon observation and experiment that is the main component of what is taught in schools. It was such science that enabled the development of nearly all aspects of modern technology. Such science enabled spacecraft to be built and launched into space, unerringly directed to reach planetary targets millions of miles away. The Mars Curiosity Rover looking for signs of life on Mars would not have happened without this kind of science. Such projects required the knowledge of Newton's laws of motion, quantum mechanics, electronics — rigorous science that combines experiment and observation with rigorous mathematical analysis. Apart from minor refinements introduced by Einstein, the validity of Newtonian physics in problems involving the motion of bodies in the solar system is beyond dispute. So also are the foundations of solid-state physics and modern electronics upon which we depend for the operation of our computers and mobile telephones. The overarching discipline that unites such rigorous science is mathematics, the propositions of which are absolute and irrefutable.

Next there is what is best described as paradigm science. This is science based on little more than pure speculation, which might be prompted by a body of empirical facts, but does not follow uniquely from the facts. The speculation involved in paradigm science is of a kind that a vast majority of people come to accept through a process that has a close analogy to the dance of bees, or the behaviour of a herd of animals. You believe in a paradigm and adopt it, not because there is overwhelming evidence in its favour, but because it is believed by the vast majority of a scientific herd. Sociology takes precedence over scientific fact.

A recent development that concerns all of us may be called internet science, which in some ways is the most objectionable and least trustworthy. Scientific debate is conducted by ill-informed bloggers who generate a confused jumble of facts, opinions and prejudices in the "blogosphere", and believe in the process that they are contributing to the development of science. This is often an unnecessary distraction, an activity that science could well do without.

It is against such a backdrop that the modern quest for the origins of life has to be assessed. Whilst technological science is yielding results that would have astounded scientists of scarcely a generation ago, *e.g.* the sequencing of the human genome, paradigm science as well as internet science impedes the progress towards understanding our ultimate origins. The present book argues that science, based on theory and experiment and verification, continues to provide ever-more compelling evidence that the currently dominant paradigm of life arising in an Earth-bound warm primordial soup is fundamentally flawed, and a new paradigm of cosmic life is long overdue. We shall argue in this book that life arose as a unique, perhaps even a near-miraculous event on a vast cosmological scale, and thereafter spread to every habitable niche in the Universe.

Chapter 1

The Genesis of Panspermia

A rigid conformity of thought would inevitably have characterised the vast majority of our pre-historic cave-dwelling ancestors. This would have been a necessary prerequisite for survival. Living as they did in small tribes, the set of rituals and beliefs of a particular tribe would have characterised that tribe, and no amount of coercion would have succeeded in affecting any change. In the 21st century, our locked-in state of paradigm fixity would appear to be a relic of such a primal instinct. Despite an avalanche of evidence pointing to a cosmic origin of life, the old Earth-centred "warm little pond" paradigm, of which we shall have much to say, seems to be deeply entrenched in our modern scientific culture and stands as an impediment to progress.

The earliest ideas in relation to our origins are buried in the mists of antiquity. In the evolution of hominids over a period of several million years the emergence of our immediate ancestor *Homo sapiens* occurred scarcely two hundred thousand years ago. Our intellectual capacity, judged by the volume of the brain, may presumably have been fixed at about this time and so also our capacity to tackle such questions as we shall discuss in the present book. From the most ancient cave paintings it could be inferred that our Paleolithic ancestors had looked up to the stars as the source of the power and mystery of life tens of thousands of years ago. All through the Egyptian, Greek and Roman times, humans seem to have vested the role of creation to deities or agencies placed in the sky.

In the line of descent of humans, the earliest fossil evidence of *Homo habilis* testifying to use of stone implements dates back to a period around 2.3 million years ago. On a geological timescale this is very recent, yet our evolutionary history extends far beyond our anthropoid ancestry to a remote past. Fossils of past life on the Earth — animals and plants — connects us in an unbroken evolutionary sequence that dates back to the great explosion of multi-celled life — the so-called metazoan explosion nearly half a billion years ago. Before this time the prevalence of unicellular life, or at any rate evidence for such life, extends quite definitely to about 3.5 billion years ago. Such evidence is in the form of stromatalites that were formed by trapping sedimentary grains in layers of biofilm laid down by cyanobacteria. Even older evidence of microbial life exists in the form of isotope anomalies connected with the relative proportions of ^{13}C and ^{12}C in the Earth's most ancient sedimentary rocks (Mojzsis *et al.*, 1996, 2001). Biology tends to concentrate the lighter isotope of carbon in preference to the heavier isotope, and this preference is detected in the most ancient rocks. Discoveries during the past decade have shown that the time of this first occurrence of terrestrial life coincides with an epoch in the Earth's geological history when impacts of comets and asteroids were at a high peak, thus strongly suggesting that the impacts themselves may have had a bearing on the emergence of the first life.

From a molecular standpoint there is an unerring unity across all life on our planet, from the smallest microorganisms to the largest and most complex plants and animals including ourselves. We all depend on the same basic chemicals. Life in all its various shapes and forms involves the interaction between two groups of complex biochemical substances, the nucleic acids and the enzymes. The nucleic acids are themselves constructed from just one sugar, four bases — guanine, adenine, thymine, and cytosine — and a phosphate, while proteins and enzymes contain about 20 separate amino acids. The myriads of possible arrangements and re-arrangements of these 26 or so basic chemical substances define the extraordinarily wide diversity of life forms that occupy our planet. We shall discover in this book that the formation of the basic chemicals poses no problem, although their precise arrangements such as in DNA and enzymes present a continuing challenge. Our thesis will be that the problem of life's ultimate origin is inextricably linked to the cosmos on the largest possible scale. Its subsequent spread and the development of diversity is a mere trivium in comparison.

Whilst humans have pondered the question of the origin of life for many millennia, all metaphysical solutions that were proposed have converged in one regard. They involve an arbitrary fiat of creation, implying an inexplicable miracle of some kind. Such a miracle may or may not involve the intervention of any named supernatural entity such as God in the Old Testament, or Allah in the Quran, or a pantheon of deities. In all cases, however, the event of creation itself was placed outside the remit of intellectual inquiry.

In ancient Greece the situation was somewhat different. The arts of dialogue and philosophy were born in classical Greece and were an important component of their tradition. To the Greeks, philosophy meant a love of knowledge and wisdom. They were concerned about the origin of things, and arising from such a quest they believed they could develop the "right way of life". Cosmological ideas — theories concerning the Universe — began to be described in the eighth and seventh centuries BC. There was a widespread belief in a "cosmic egg" from which all things were thought to have emerged. By the end of the seventh century BC such ideas became encapsulated in the Homeric legends and came to be established as an integral part of the Greek culture.

Thales of Miletus (c. 624–546BC) was perhaps the first to generalise questions relating to origins and to attempt rational explanations that did not involve the Gods. Thus began the long process to separate theology and science, a process that continues even to the present day. Thales observed that water was the most abundant material on Earth and that all plants and animals needed it for survival. He was thus driven to propose that life originated from water.

The materialistic ideas concerning the world that were proposed by Thales reached their summit with Democritus of Abdera (470–380 BC). Democritus thought that the essential property of life is the possession of a soul or *psyche*. He believed that the Universe was made of atoms that moved in space and that all physical change involved arrangements and rearrangements of these atoms. The existence of life was contingent on these atoms, and death came when life-giving atoms departed from the body to take residence in the sky.

The doctrine of spontaneous generation was probably first proposed by the pre-Socratic philosopher Anaximander (611–547 BC). He stated that all living creatures spontaneously originated in "wet" soil (mud) when acted upon by the Sun, and, moreover, that the earliest creatures differed from those that exist today. Thus Anaximander might arguably be reckoned as one of the earliest philosophers to propose an evolutionary theory of life. Because of the great length of infancy in humans he argued that they must have originated earlier in some form endowed with a higher level of independence and maturity, perhaps as a type of fish; otherwise he argued they would not have survived. Anaximander further claimed that the process of spontaneous generation of life must even continue to the present day, with aquatic creatures emerging directly from inanimate matter. Anaximander's disciple, Anaxagores of Clazomenae (570–500 B.C.) shed a somewhat different light on the subject. He argued that it was not water but air that was the primary life-giving agent. Air enveloped the Earth and permeated the bodies of all living things.

Neither Anaximander nor Anaxagoras had much influence on the course of history, however, compared with Aristotle (384–322 BC), who took centre-stage two centuries later, and laid the foundations of the whole of Western natural philosophy. Aristotle adopted Anaximander's doctrine of spontaneous generation claiming moreover that he had firm evidence to support it. Of the several pieces of evidence cited in his writings, one example goes thus:

"This (spontaneous generation) occurs in ponds, especially one near Knidos, which, so it is said, on one occasion dried up at the time of the Dogstar and all the mud was taken out; water began to collect in it as the first rains came, and at that point tiny fishes appeared …… This evidence shows that certain fishes are produced spontaneously, and do not come out of eggs or from copulation."

… and another runs thus:

"Now there is one property that animals are found to have in common with plants. For some plants are generated from the seed of plants, whilst other plants are self-generated through the formation of some elemental principle similar to a seed; and of these latter plants some derive their nutriment from the ground, whilst others grow inside other plants…. So with animals, some spring from parent animals according to their kind, whilst others grow spontaneously and not from kindred stock; and of these instances of spontaneous generation some come from putrefying earth or vegetable matter….."

(*From* Peck, A. L., 1970. Aristotle: *HistoriaAnimalium*, Books IV–VI. Harvard University Press)

Of course the widely quoted and most enduring example of spontaneous generation is of Aristotle's observation that "fireflies emerge from a mixture of warm Earth and morning dew..."

Although it was Aristotle's empirical approach that evidently led him to the theory of spontaneous generation of life, his observations at the time were necessarily too superficial and of course flawed. Why, one might ask, did he not opt for a more plausible explanation of how certain animals come into existence, such as the idea that seeds, eggs, or larvae could persist through harsh circumstances and be carried across great distances? From a philosophical standpoint such a position may have been seen to be at odds in general terms with an Earth-centred Aristotelian Universe which was the dominant world view at the time.

A few decades after Aristotle, the Greek astronomer and mathematician Aristarchus of Samos (310–230 BC) introduced the idea of πανσπεμια (panspermia) — seeds everywhere, thus rejecting the doctrine of spontaneous generation. Aristarchus used observation and measurement to calculate the correct distances to the Sun and to the Moon, and he was thus forced to reject Aristotelian Earth-centred cosmology. The world, however, was not ready for either a heliocentric system or for Aristarchus' ideas of panspermia. Heliocentric cosmology took another 1700 years to get accepted, and as we shall see in this book, panspermia is only now slowly drifting into the domain of respectable science.

Spontaneous generation took many different forms throughout its long history. It was augmented along the way by an additional principle of vitalism — implying the existence of a life force (spirit or *pneuma*) distinct from the inanimate matter from which life was supposed to spring. The Roman physician Galen of Pergamon (Claudius Galenus, AD 129–216) first introduced the idea of *pneuma* and believed that we inhale this life-giving entity from the air. In the fifteenth and sixteenth centuries the concept of *pneuma* even influenced the development of physics, helping to shape theories of aether and to accommodate Newton's ideas of action at a distance. René Descartes (1596–1650) sharpened earlier distinctions of vitalists between mind and matter — the philosophy of Cartesian dualism as it has come to be known. The separation of a non-living and living world was emphasised by followers of vitalism, and in this context they maintained that organic chemical characteristic of life could not be made synthetically by inorganic processes.

When Friedrich Wohler successfully synthesised urea from inorganic chemicals in 1826, vitalists suffered a setback. Vitalism fell out of favour but nevertheless continued to have vocal advocates even into the twentieth century. The most noteworthy amongst them was Hans Driesch (1867–1941), an eminent embryologist, who explained the life of an organism in terms of the presence of an *entelechy*, an entity controlling organic processes. Likewise, the French philosopher Henri Bergson (1874–1948) posited an *élan* that was required to overcome an inherent resistance of inert matter to form into living things.

Attempts to either re-affirm or disprove Aristotle's doctrine of spontaneous generation have punctuated the history of science for many decades. In 1668 an Italian physician Francesco Redi came up with an alleged disproof of spontaneous generation — specifically, the proposition that maggots arise spontaneously from rotting meat. He had shown that whatever gave rise to the maggots must have travelled to the meat through the air, and when such ingress was prevented maggots did not appear. Between 1745 and 1748, John Needham observed that microorganisms flourished in various soups and broths that had been exposed to the air. On this evidence he claimed that there was a "life force" in all the molecules of inorganic matter, including air, that could lead to the creation of life.

Two decades later, Lazzaro Spallanzani, an Italian abbot and biologist, repeated Needham's soup experiments with much better controls, boiling and using closed and open containers, and he concluded that the microorganisms in the flasks containing the spoiled soup must have entered from the air. A bitter argument developed between Needham and Spallanzani concerning sterilisation procedures, with the claim by Needham that Spallanzani's over-boiling had killed the "life force" and that bacteria could not develop by spontaneous generation in the sealed containers because new "life force" could not enter from the air. All this may sound a little whimsical now, but at the time they would have represented the summit of intellectual inquiry and discourse.

Jean-Baptiste Lamarck (1744–1829) proposed his theory of inheritance, in order to reconcile his ideas with the still fashionable doctrine of spontaneous generation. He argued that as creatures strived for greater adaptation and moved up on the ladder of complexity, new organisms arose *de novo* to fill the vacant gaps on the lower rungs.

By 1860, this controversy had become so heated that the French Academy offered a prize for any experiment that would help resolve the long-standing dispute once and for all. Louis Pasteur carried out an experiment using a special flask which had a curved neck permitting air in, but preventing any microorganisms from entering a nutrient broth. Pasteur showed that the broth in this flask remained free from microorganisms, even though it was open to the air. By this experiment he concluded that microbes could only come from other microbes, and for this discovery Pasteur was awarded the coveted prize. Pasteur then went on to generalise his conclusion to all life: "*Omne vivum ex vivo*" — "all life from life", and it was this pronouncement that was to have the most profound implications for biology and indeed the whole of science.

Pasteur's life-from-life dictum implied that each generation of every plant or animal is preceded by an earlier generation of the same plant or animal. This view was taken up enthusiastically by others, particularly by physicists, among whom John Tyndall lectured frequently on the London scene, as for instance in a Friday evening discourse at the Royal Institution on 21 January 1870. It was to this lecture that the editorial columns of the newly established journal *Nature* objected with some passion. Behind the objection was the realisation that was Pasteur's claim taken to be strictly true, the origin of life would need to be external to the Earth. For if life had no spontaneous origin, it would be possible to follow any animal generation by generation back to a time before the Earth itself existed, the origin being therefore required to have taken place outside the Earth.

This was put in remarkably clear terms in 1874 by the German physicist Hermann von Helmholtz (in *Handbuch der Theoretische Physik*, Vol 1, transl. H. von Helmholtz and G. Wertheim, Braunschweig, 1874):

"It appears to me to be a fully correct scientific procedure, if all our attempts fail to cause the production of organisms from non-living matter, to raise the question whether life has ever arisen, whether it is not just as old as matter itself and whether seeds have not been carried from one planet to another and have developed everywhere where they have fallen on fertile soil ..."

Sir William Thomson (Lord Kelvin) also said of Pasteur's paradigm:

"Dead matter cannot become living without coming under the influence of matter previously alive. This seems to me as sure a teaching of science as the law of gravitation ..."

So if life had preceded the Earth, how had it arrived here and where had it come from? Earlier in the 18th century the German physician R.E. Richter had suggested that living cells might travel from planet to planet inside meteorites. Inadequacies in Richter's dynamics permitted J. Zollner in the 1870's to raise objections, eagerly seized on by orthodox opinion. But Kelvin's superior knowledge of dynamics allowed him to see that there was nothing to Zollner's objections, in particular that evaporation from the outside of a large meteorite and its low thermal conductivity keeps its interior cool, thereby reasserting the possibility of organisms being carried from planet to planet inside meteorites. In his presidential address to the 1881 meeting of the British Association, Kelvin drew the following remarkable picture.

"When two great masses come into collision in space, it is certain that a large part of each is melted, but it seems also quite certain that in many cases a large quantity of debris must be shot forth in all directions, much of which may have experienced no greater violence than individual pieces of rock experience in a landslip or in blasting by gunpowder. Should the time when this earth comes into collision with another body, comparable in dimensions to itself, be when it is still clothed as at present with vegetation, many great and small fragments carrying seeds of living plants and animals would undoubtedly be scattered through space. Hence, and because we all confidently believe that there are at present, and have been from time immemorial, many worlds of life besides our own, we must regard it as probable in the highest degree that there are countless seed-bearing meteoric stones moving about through space. If at the present instant no life existed upon the earth, one such stone falling upon it might, by what we blindly call natural causes, lead to its becoming covered with vegetation."

Thus over 133 years ago the ideas we shall develop in this book were already well-developed. It is a feature of scientific method that unless an idea has a means of advancing itself through observation or experiment it stultifies, almost regardless of how good the idea may be in itself. Unfortunately there was no way at that early date, 1881, whereby observation or experiment could be brought seriously to bear on Kelvin's formulation of panspermia. The experiments related to the possibility of meteorites harbouring viable life have been possible only relatively recently as we shall see in this book.

Historically the next facet in the story was associated with Svante Arrhenius, a Nobel Prize winning Chemist, whose book *Worlds in the Making* appeared in English, published by Harpers of London, in 1908 (Arrhenius, 1903, 1908). Arrhenius' book pointed out that microorganisms possessed the right range of sizes for them to be propelled by the pressure of starlight from one planetary system to another. He also argued that microorganisms possess unearthly properties, properties which cannot be explained by natural selection against a terrestrial environment. The example for which he was responsible was the taking of seeds down to temperatures close to absolute zero, and of then demonstrating their viability when reheated with sufficient care. Many other unworldly properties have since been discovered, as for instance the ability of microorganisms to survive inside a nuclear reactor. In the past two decades microbiological research into microbes called extremophiles have taken this early work to new dimensions, with the unravelling of survival attributes that exceed all previous expectations, and were certainly unknown to Arrhenius. These include survival at high and low temperatures, high acidity and high alkalinity, at depths in the Earth's crust in total darkness, high in the stratosphere — to name but a few. It is difficult to imagine how such an amazing set of properties arose in the context of the Earth, but they all highly relevant to survival in space, and all point to the relevance of panspermia.

Chapter 2

The Primordial Soup and Evolution

After the setback to the doctrine of spontaneous generation following Pasteur's experiments that was discussed in Chapter 1, the question of the origin of life from inanimate matter was taken up again in the twentieth century. The need for an empirical approach to this problem was recognised and a range of theories came to be formulated and discussed.

The Russian scientist A. I. Oparin and the English biologist J. B. S. Haldane first recognised that life must ultimately have an inorganic chemical ancestor and they proposed an explicit model on this basis (Oparin, 1924, 1938; Haldane, 1928, 1954). The general belief at the time was that there could be no organic molecules indigenous to the primitive Earth before the inception of life. The starting chemical system was accordingly taken to be a mixture of simple inorganic gases (molecular hydrogen, methane, ammonia, water) that was supposed to be present in primordial atmospheric clouds. Such stable molecules, by simply colliding with each other in the gas phase, cannot form complex organic molecules. So ultraviolet light from the Sun and energetic processes like thunderstorms and electrical discharges were invoked as a first step to break up the postulated inorganic molecules into 'energized' fragments or radicals. The 'energized' fragments (radicals) then react and recombine in a cascade of chemical reactions and in this process a small yield of prebiotic molecules are produced. These molecules rain down into the primitive oceans to form the primordial soup in which the origin of life is supposed to have occurred.

As far as the formation of organic molecules is concerned the proposed scheme cannot be questioned provided the starting mixture of atmospheric gases was of a type that was proposed, that is to say, chemically non-oxidizing or reducing in character. This latter assumption was subsequently found to be incorrect. The early Earth's atmosphere inferred from the states of oxidation found in the oldest rocks, was discovered to have been oxidizing rather than reducing. This oxidizing property is unfavourable for the persistence of complex organic molecules, and no significant yields of biochemicals could form under these conditions. The primordial organic soup theory in its original form as proposed by Oparin and Haldane was therefore discovered to be inappropriate to the real Earth.

However, long before such facts relating to the early atmosphere came to be discovered, several laboratory experiments were conducted with a view to testing the validity of the Oparin–Haldane model. In 1953 the American chemist Stanley Miller showed that high yields (up to three percent) of amino acids could be formed by sparking a mixture of molecular hydrogen (H_2), methane (CH_4), water (H_2O) and ammonia (NH_3) continuously for one week (Miller, 1953; Miller and Urey, 1959). Later Cyril Ponnamperuma and others successfully extracted traces (about 0.1 percent) of the components of nucleic acids after irradiating similar mixtures with an electron beam (Ponnamperuma and Mack, 1965). They also produced trace quantities of sugars in other experiments. Thus the key organic monomers of life were shown to be produced from inorganic molecules in the laboratory with relatively little difficulty. These results were undoubted triumphs for experimental chemistry, but they did not constitute proof of the primordial soup theory of origin of life as usually claimed. The building blocks of life — amino acids, nucleotides, sugars — still remained a far cry from life.

In a modern variant of the Oparin–Haldane primordial soup model, geothermal vents, rather than the oceans, lakes or ponds, have been considered as appropriate settings for a terrestrial origin of life. Whilst the organic molecules needed for life could be produced fairly easily in a multiplicity of settings, the transition from such organic building blocks of life to the simplest self-replicating living cell involves what might be a nearly insuperable improbability hurdle (Hoyle and Wickramasinghe, 1980, 1982). In uncritically accepting such models for the origin of life scientists may have unwittingly replaced the mysteries that shrouded this question in earlier times with equally intractable scientific dogmas.

It is generally conceded that the path from chemicals in a hypothetical primordial soup, in whatever location, to a self-replicating living cell must progress along a sequence of organisational steps of ever-increasing complexity. The essence of life at the molecular level is undoubtedly the highly specific arrangements of certain monomers into long chain polymers — amino acids into proteins and nucleotides into nucleic acids (DNA or RNA). In other words, the highly specific information content of these long-chain polymers essentially defines life.

Proteins are the macromolecules responsible for structure of living cells as well as their function. They form an important component of a cell wall structure within which the biochemistry of life is contained. Enzymes are the subset of proteins that catalyse all the crucial chemical processes of life. DNA and RNA are the molecules that carry the information content of life and serve as the vehicle of heredity. DNA and proteins can thus be seen as complimentary parts of every living system. The origin of the most primitive life form from the monomers — amino acids, nucleotides — must somehow incorporate these two complimentary parts within a single cell-bound system that is capable of metabolizing, harnessing energy from the environment, replicating, evolving and ultimately generating the whole spectrum of life.

The most popular contender for a first step in this evolutionary process is the so-called RNA-world where dual roles of function (catalysis) and heredity are combined in a single RNA (ribozyme) macromolecule. Here nucleotides polymerize into RNA chains that serve as autonomously replicating and catalysing macromolecules without the need for any intermediary enzyme system (Orgel and Crick, 1968). Likewise, other contenders for the initial stage of prebiotic development that have been suggested include the "Iron-sulphur world" (Wachtershauser, 1990), the "PNA (peptide nucleic acid) world" (Bohler *et al.*, 1995) and the "Clay World" (Cairns-Smith, 1966).

The iron-sulphur world requires an organic soup in the form of water (steam) with dissolved gases CO, NH_3, H_2S such as would exist in volcanic outflows coming in contact with mineral catalysts like iron sulphide or nickel sulphide. In the PNA world highly robust chains of peptide nucleic acid (PNA) replace the RNA in the RNA-world models. In the clay-world theory complex organic molecules become ordered on clay lattice surface structures that can hold patterns so as to act as primitive inorganic templates. In view of the relatively high cosmic abundance of silicon, the latter clay-world model might well have a special role to play in the origin of life in a cosmic context.

The transition from any one of these intermediate systems to a protogene system possessing prescriptive information for evolution, and finally to DNA-protein-based cellular life is still in the realm of pure speculation (Abel and Trevors, 2006). An unsolved problem is the origin of the genetic code by which the 4 RNA bases arranged in triplets correspond to amino acids in enzymes. In a later chapter (Chapter 11) we shall discuss the argument that the genetic code is not itself arbitrary, and that it may even be used to transmit an intelligent message (SETI) in a coded form.

Once cellular life has somehow originated, the functioning of even the most primitive living system depends on thousands of chemical reactions taking place within a membrane-bound cellular structure. Such reactions, grouped into metabolic pathways, have the ability to harness chemical energy from the surrounding medium in a series of very small steps, transporting small molecules into the cells, building biopolymers of various sorts, and ultimately making copies of itself possessing a capacity to evolve. Batteries of enzymes, comprised of chains of amino acids, play a crucial role as catalysts precisely controlling the rates of chemical reactions. Without such enzymes there could of course be no life.

From the turn of the 20th century the important metabolic pathways in biology have been systematically unravelled — *e.g.* the CO_2 cycle in plants, the citric acid cycle (Kreb's cycle) *etc*. However, even if we were to gain a complete knowledge of all the metabolic pathways in biology, we would not, in the opinion of the present author, come any closer to comprehending how, when and where the first living system emerged.

As mentioned before, in present-day biology the information contained in the enzymes is crucial for the operation of life, and this information is transmitted via the coded ordering of nucleotide triplets in DNA. In a hypothetical RNA-world that may have predated the DNA-protein world, RNA may have served a dual role as both enzyme and genetic transmitter. If a few ribozymes are regarded as being the precursors of all life that was later to evolve, one might attempt to make an estimate of the probability of assembly of a simple ribozyme comprised of say 300 bases. With the 4 coding bases of nucleic acid chains this probability turns out to be 1 in 4^{300}, equivalent to 1 in 10^{180}, which can scarcely be supposed to happen even once in the entire 13.8-billion-year history of the Universe.

An analogous argument at the level of gene evolution has led to a similar conclusion. The smallest known autonomous bacterium *Mycoplasma genitalium* has approximately 500 genes. If this set of genes is to be generated in the context of a terrestrial primordial soup, the odds against this happening are superastronomical. A metaphor used by Fred Hoyle to emphasise this improbability is to liken such an occurrence to a "tornado blowing through a junk yard assembling a Boeing 747". Hoyle and the present writer accordingly developed the theory of cometary panspermia in which the origin of life in the form of a "pangenome" required a physical system that transcends the scale of our planetary system by very many orders of magnitude. The origin of the pangenome in this picture is a unique "Big Bang like" creative cosmological event, and its subsequent dispersal and reassembly in countless locations becomes a cosmic imperative. A similar argument was invoked by Crick and Orgel (1973) in their proposal of "directed panspermia" in which an artificially engineered life system is deliberately imposed on the Earth. Darwinian evolution leading to a succession of small adaptive changes of species caused by mutations and natural selection becomes in these models a fine-tuning process, which although important in carving out details, is not the primary process of evolution. In the view that we shall explore in this book, an initially sterile Earth was infected by the arrival of pre-evolved genetic packets contained within bacteria and viruses. As we shall see in later chapters these are the pangenome components that were generated over very long time spans on a cosmological scale and delivered by the agency of comets.

There have been many ambitious schemes to attempt a laboratory simulation of the origin of life. In one such scheme David Deamer generated trillions of half-micron-sized cellular compartments (bubbles) by adding water to dry lipids (fats) in a flask (Deamer, 2011). To this is added a solution of small peptides and nucleic acids, in the hope that amongst the trillion or so cellular compartments, and a vast array of combinatorial arrangements of the biological monomers, a proto-living system will be found. The failure to witness any trend whatsoever towards the emergence of a living system has been attributed to the infinitesimal scale of the laboratory system when compared to the terrestrial setting in which life is thought to arise. Yet, if we move from the laboratory flask to all the oceans of the Earth we gain in volume only a factor of $\sim 10^{20}$, and in time from weeks in the laboratory to perhaps half a billion years, the gain is a factor of 10^{10}. In the probability calculation for the single ribozyme we thus gain only a factor of 10^{30} in all, reducing the improbability factor from 1 in 10^{180} to 1 in 10^{150}. On this basis it is very difficult to avoid the conclusion that the emergence of the first evolvable cellular life form was a unique event in the cosmos as we shall argue in a later chapter. If this unique event did indeed happen on Earth for the first time, it must be regarded as a "miraculous" event that is unlikely to be repeated elsewhere, let alone in any laboratory simulation of the process.

The difficulty of finding unequivocal evidence of the relics of prebiology in the geological record could be seen as another problem for Earth-based theories of the origin of life. The suite of organics present in interstellar clouds consistently directs us to possible origins away from Earth to more and more distant parts of the Universe. At the very least the organic molecules needed for life's origins are much more likely to have been generated in a cosmic context rather than being formed *in situ* on Earth. This was indeed the first "weak" version of panspermia proposed by Fred Hoyle and the author in 1976 (Hoyle and Wickramasinghe, 1976, 1978). We argued, based on astronomical data, that it was more likely that organic molecules that were known to be present in interstellar clouds and comets were delivered to Earth by comet impacts. It is now becoming amply clear that life arose on Earth almost at the very first moment that it could have survived at the end of a long period of comet impacts on the Earth 4.2–3.82 billion years ago. The proposition that impacting comets carried the first life to the Earth from the external Universe cannot be excluded on this basis.

If one accepts the calculations discussed above showing vanishingly small *a priori* probabilities for the transition of non-life to life, it would appear that two options remain:

(a) The origin of life on Earth was an extremely improbable event that occurred against all the odds and is unlikely to be reproduced elsewhere. Life might then be thought of as unique to Earth, placing our planet once again in a privileged position in the cosmos.

(b) The alternative is that a very much bigger physical system than was available on Earth and perhaps a much longer timescale were needed for the initial event of origination. It was from such a cosmic origin that life was transferred to Earth.

How big or how old that cosmic system needs to be is still a matter for debate. Arguments by Abel and Trevors (2006) and Abel (2009) suggest that within the framework of Big Bang type cosmologies naturalistic protogene formation still faces almost insuperable impediments. The situation is more favourable in a class of quasi-steady-state cosmologies in which much longer timescales and larger cosmological volumes are available (Hoyle, Burbidge and Narlikar, 2001). However, by whatever process life emerged, the event of cosmic biogenesis must be reckoned as unique and its subsequent spread throughout the Universe assured by the processes of "panspermia" (Hoyle and Wickramasinghe, 2000).

The argument that panspermia must be rejected because it merely transfers the problem of origin from Earth to another setting is by no means scientific. The question of whether life started *de novo* on Earth, or was introduced from the wider Universe, is a fully scientific question that merits investigation, and one that is open to test and verification. The invocation of Ockam's razor to exclude a discussion of such matters is also inappropriate. It is an excuse for keeping scientific discussion within the strict bounds of what is currently considered "orthodox", and is strikingly reminiscent of very similar strictures that stifled science throughout the Middle Ages.

Whilst the ultimate origin of life remains in the realm of speculative science, evolution itself is an indisputable empirical fact. The unsolved problems relate to the cosmic scale over which evolution occurred. The fossil record shows clear evidence for the progression of life forms on Earth from simple, single-celled microorganisms four billion years ago to the entire panorama of life that exists today. Life forms have evolved through geological time, changing and branching out in a myriad ways to lead eventually to the totality of life on our planet. Hundreds of millions of fossils and microfossils are preserved in well-dated sequences of sedimentary rocks. The evidence is clear that the planet-wide distribution of species has changed over time alongside with the changes that have occurred in the physical and chemical conditions of the Earth's surface and atmosphere.

The precise mechanism by which species have evolved from simple to complex forms remains open to argument. Charles Darwin and Alfred Russell Wallace proposed a mechanism involving natural selection — survival of the fittest (Darwin and Wallace, 1858). In its modern form this depends on the random genetic mutations that inevitably accompany reproduction and the vast numbers of progeny to select for best adaptations to ever-changing environments. The idea is that minute changes in a species produced by mutations and natural selection build up over very many generations until a new species finally emerges. In this process, if it were confined to our planet, the expectation will be to see a continuous record of variation of species in the terrestrial fossil record, and a preponderance of "intermediate forms" which, as it turned out, are difficult to find. Darwin himself worried about this and wrote thus in his "Origins of Species":

"Geology assuredly does not reveal any such finely graduated organic chain; and this, perhaps, is the most serious objection that can be urged against the theory…"

Darwin's consolation at the time was that the fossil record was imperfect and incomplete, but this has since been rectified by later discoveries of fossils and also of molecular biology, which still do not show the dominance of intermediate forms. As we shall see later this objection disappears if Darwinian evolution is not a process that is confined to the Earth. If the Earth receives periodic injections of new genetic information from the cosmos, transitional forms may have only a brief and undetectable presence in the record of terrestrial fossils. The problem now is not so much gaps in the fossil record but the huge surges of evolution that characterise the record of terrestrial biology. Such surges of evolution and sporadic injections of diversity punctuate an otherwise uneventful record of evolutionary progress over hundreds of millions of years.

The concept of "punctuated equilibrium" was introduced by Stephen Jay Gould and Niles Eldridge to explain the available data in a phenomenological way (Gould and Eldridge, 1977). The facts clearly show that species distributions remain relatively static and stable over very long periods, and this stability is abruptly interrupted by surges of speciation and extinction. The genetic mechanism by which entirely new orders or classes suddenly arise is still somewhat obscure. The existing theories that confine the entire process of evolution to a closed system on the Earth face insuperable problems. These problems disappear, however, if evolution is considered to be a galaxy-wide or a cosmic phenomenon. The progression of life on the Earth will then be the result of selecting from an almost infinite range of genetic possibilities that are on offer from the Universe at large. New genes reflecting an evolutionary history spanning cosmologically vast timescales and distances will be on offer. The emergence of new orders, classes or even phyla could be explained as the result of the ingress of new and innovative genes. The details of such a process will be discussed in a later chapter.

Whatever the precise mechanism might be, evolutionary progress over geological time can be represented by "trees of life" or phylogenetic trees that seek to connect different species. Schemes of this type predated Darwin, although it was Darwin who re-introduced the idea to explicitly depict the progress of evolution.

More elaborate and detailed phylogenetic trees were developed with the advent of molecular biology, where random mutations in highly conserved genes are used as molecular clocks. For instance the number of amino acid differences in molecule cytochrome C from a bacterium to mammals through a swathe of intermediate organisms defines a time chart of genetic evolution spanning billions of years. Similarly for higher life forms changes in molecules like haemoglobin have been widely used as an index of species divergence.

The form of this tree of life that is widely accepted is due to Carl Woese and is based mainly on changes in rRNA genetic sequences (Woese, 1967; Woese and Fox, 1977). Three domains of life — Bacteria, Archaea and Eucarya (Fig. 2.1) have been clearly separated in this way.

The gradual changes and connections implied through the branches of this tree have prompted biochemists to look for a "root" of the tree of life — the Last Universal Common Ancestor (LUCA). This search has been foiled, however, by the circumstance that the genes used for tracing evolutionary progress in all three domains (branches) have very often been transferred across the branches.

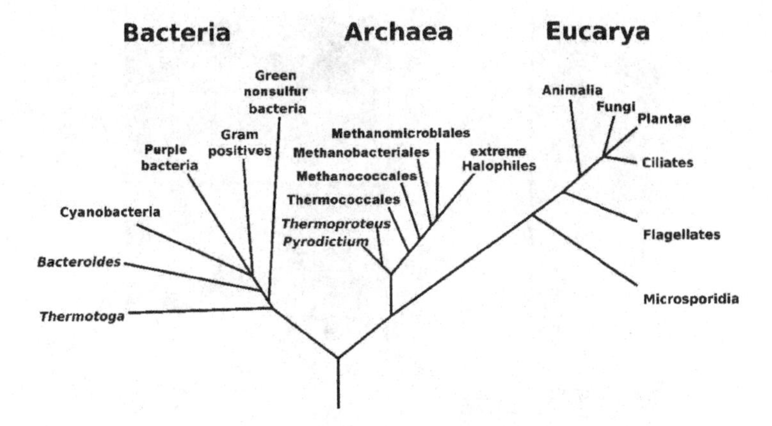

Fig. 2.1. Three domains of life.

As we go nearer to the root of the tree of life, connections determined by tracking changes in crucial genes are turning out to be exceedingly fuzzy, and many investigators have decided that there might not be a single last universal common ancestor. This has led to a transfer of interest from "LUCA" to the concept of a "pangenome", which can be envisaged as a primordial ensemble of genes. We shall argue in this book that this entire ensemble of genes is most likely to be rooted in the cosmos, and the genes needed for terrestrial evolution continue to be derived gratis from the cosmos.

Apart from the extended periods of "stasis" in the evolutionary record that we already mentioned there is an even more serious problem for any model of evolution confined to Earth. This is the fact that ancient microorganisms do not differ significantly either morphologically or genetically from their modern counterparts. Schopf and his collaborators have discovered fossil bacterial forms dating back to 3.5 billion years that are morphologically indistinguishable from modern cyanobacteria (Fig. 2.2) (Schopf, 1999, 2006). It has also been found that microorganisms recovered from the abdomens of 30–220 Myr old bees and other insects trapped in amber show identical morphologies to modern microbes (Cano and Borucki, 1995). Vreeland *et al.* (2000) obtained a similar result for even older microbes in salt crystals from a New Mexico salt mine that date back to 250 million years, and show furthermore that a Bacillus recovered from these ancient salt crystals is genetically indistinguishable from a modern Bacillus. Such discoveries

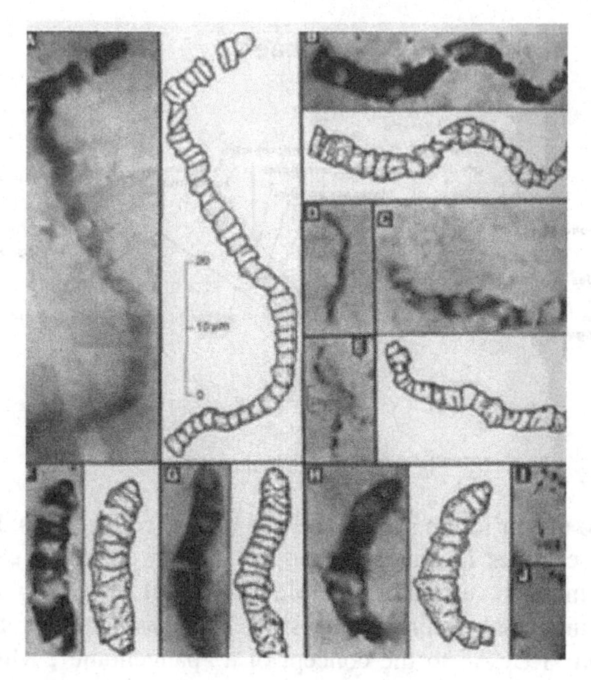

Fig. 2.2. Microfossils of cyanobacteria in Precambrian sediments (Schopf, 1999).

point to a continual replenishment of pristine microbes and microbial genes that outnumber and overwhelm an older population in which mutational changes have built up.

It cannot be overemphasised that a fundamental error of conventional Darwinism is to constrain evolution to be Earth-centred, occurring only at the molecular level of individual DNA codons, rather than at the genetic level involving the introduction of entirely new genes. Evidence over the past few decades has shown the closed-box evolution picture to be fatally flawed.

Chapter 3

Cosmological Context

In view of the superastronomical improbabilities involved in the emergence of the simplest self-replicating living cell, it stands to common sense that it will pay handsomely to go to the largest available system to solve the problem of our origins. And the largest available system, by definition, is the entire Universe, thus leading us inevitably to cosmology. Before discussing what cosmological options are available, let us first note that the most crucial chemical element for life is carbon. Without carbon there can be no life as we know it, so carbon must be present extensively in the Universe before life can emerge. In addition to carbon life requires other elements, H, N, O, P, S, Mg amongst others.

According to so-called "conventional" cosmological models, most if not all the material we observe in the Universe was generated in a singularity known as the "Big Bang", believed to have occurred 13.8 billion years ago. Within the first 20 minutes of this event, Big Bang nucleosynthesis is complete, resulting in the synthesis of hydrogen, helium, and trace amounts of other elements — lithium, boron and beryllium. The production of all other chemical elements in the Universe takes place in the deep interiors of stars.

By the mid-1930's it was already known that stars, during most of their lives, derive their energy by the fusion of hydrogen to helium which takes place at temperatures of the order of a few million degrees Kelvin. This is the nuclear reaction that keeps the Sun and most other stars shining in the night sky for most of their lives. It provides the energy for life on Earth and possibly on other planets as well.

At the present time the Sun has expended about half its initial store of hydrogen fuel, and in another 5 billion years or so the hydrogen will be completely exhausted. What happens after all the available H in a star is transformed into He forms an important part of our story. The next logical step in the chain of energy-producing nuclear reactions is a reaction that combines three nuclei of helium-4 into a nucleus of carbon-12. But this was considered in the 1930's to be impossible and prohibited according to what was known at the time about nuclear physics and in particular the structure of the carbon nucleus. Fred Hoyle came in at this stage (in 1946) to argue that the nuclear reaction (known as the triple-alpha reaction) that was thought to be forbidden, *must* take place if carbon is to be produced in any quantity in the Universe as it is indeed required for our very existence. By this assertion he unwittingly inaugurated what has come to be known as the anthropic principle. Our existence in the Universe and the consequent requirement for the widespread availability of carbon demand certain things to be true. One of these was a firm imperative for the nucleus of the carbon atom to possess a short-lived excited state (the Hoyle state) at an energy of 7.75 MeV. Hoyle predicted the existence of this state theoretically and suggested to nuclear physicist W.A. Fowler that he should look for it in his laboratory at Caltech. Reluctantly and after much persuasion Fowler searched for this state at the Kellog Radiation Laboratory at Caltech and indeed found it. For this discovery which transformed astronomy he was awarded the Nobel Prize for Physics in 1983.

With the discoveries of Hoyle and Fowler in the 1950's the sequence of nuclear transformation in stars from hydrogen to helium to carbon and thence to the entire suite of elements in the periodic table was completed. All the chemical elements essential for life could now be understood as the products of nucleosynthesis in stars, and they are scattered throughout interstellar space by the agency of exploding supernovae and the outflows of dust from stars and star systems.

The basic molecules that may be required for any putative prebiotic chemistry — including H_2O, simple organics, including polyaromatic hydrocarbons PAH's — are known to be present in vast quantity in our galaxy (see Wickramasinghe *et al.*, 2010). We shall argue later that a large fraction of all these molecules are most likely to be the degradation products of biological structures such as bacteria and viruses that were generated cosmologically and stored within comets. Such objects, replete with the legacy of cosmic life, become involved in every star and planet forming event in the disc of our galaxy and indeed of galaxies in general. If one adopts such a model the build-up from simple to complex organic molecules and prebiotic evolution in interstellar space becomes largely irrelevant.

In the standard picture of prebiotic evolution, biochemistry is supposed to start from simple organic molecules that form *in situ* in space. In such a scheme all one could realistically hope to achieve in the way of progress towards biochemistry in interstellar clouds is the production of moderately complex organic molecules by means of gas-phase chemistry or surface chemistry on dust grains. The more complex of the organic molecules so formed must then enter a watery medium in suitably high concentrations to begin the presumptive prebiotic chemistry that may have eventually led to life. The situation is essentially a revival of the primordial soup theory in a different context, one that inherits all the probability obstacles we discussed earlier.

In a galaxy such as our Milky Way, stars and planets form from clouds of gas and dust, and in the case of our solar system, the first solid bodies to condense were the comets. These icy objects would have mopped up the organic molecules and dust of the parent interstellar cloud, and for a few million years after they condensed would have possessed bio-friendly liquid water interior domains, due to the heating effect of radioactive decay of the short-lived nuclide ^{26}Al. If even the minutest amount of microbial life was present in dormant form in the parent interstellar cloud, the newly-formed comets that contain these living cells will amplify them on a very short timescale. We have already mentioned that the start of life on the Earth coincides with an epoch of comet impacts pointing strongly to comets being the source of terrestrial life.

To begin with let us consider a prospective site of origin of life in a purely galactic context. Prior to life being generated anywhere in the galaxy, comets, heated by the energy of radioactive decays, would have provided trillions of “warm little ponds” replete with water and organic nutrients, and their huge numbers could have diminished significantly the improbability hurdle for life to originate in one of them. Recent studies of Comet Tempel 1 have shown evidence of organic molecules including PAH’s, clay particles as well as liquid water in comets, providing an ideal setting for the operation of the “clay theory” of the origin of life (Wickramasinghe *et al.*, 2010).

A single primordial comet of this kind will be favoured over all the shallow ponds and edges of oceans on Earth by a factor 10^4, if we take into account the total clay surface area for catalytic reactions as well as the timescale of persistence of each favourable candidate location. With 10^{11} comets in the Oort cloud, the factor favouring solar system comets over the totality of terrestrial "warm little ponds" weighs in at a figure of 10^{15}, and with 10^{10} Sun-like stars replete with comets in the entire galaxy we estimate a factor of 10^{25} in favour of a cometary origin of life.

The next step in the argument is that once life got started in some comet somewhere, its spread in the galaxy becomes inevitable. The comets themselves serve as the amplifiers and distributers of life in the galaxy. Dormant microorganisms are released in the dust tails of comets and propelled by the pressure of starlight to reach interstellar clouds. When a planetary system forms the newly-condensed comets in that system provide sites for the amplification of surviving microorganisms that are incorporated in the new system. The transport of microorganisms and spores within the frozen interiors of comets carries only a negligible risk of destruction — they are essentially immortal. But transport of microbes in either naked form, within clumps of dust or within meteorites entails varying degrees of risk of inactivation by cosmic rays and UV light. However the successful seeding of life requires only the minutest survival fraction between successive amplification sites, and even partially corrupted or inactivated genetic messages in bacteria and viruses may be adequate to transfer the crucial information for life (Wickramasinghe, 2011; Wesson, 2010). Of the bacterial particles included in every newly-formed planetary system we can estimate that only one in 10^{24} needs to remain viable to ensure a positive feedback loop for panspermia.

All the indications are that this is a very modest requirement that is hard, if not impossible, to violate. The situation is analogous to the sowing of seeds in the wind. Few are destined to survive, but so many are the seeds that some amongst them always take root.

Whilst comets could provide a more or less continuous supply of primitive life (archeae and bacteria) to interstellar clouds and thence to new planetary systems, the genetic products of evolved life on planets like Earth could also be disseminated on a galaxy-wide scale. This process also involves comets by a mechanism whereby impacts splash back life-laden dust into space. Our present-day solar system, which is surrounded by an extended halo of some 100 billion comets (the Oort cloud), moves around the centre of the galaxy once every 240 Myr. Every 40 million years, on the average, the comet cloud becomes perturbed due to the close passage by a molecular cloud. Gravitational interactions lead to hundreds of comets from the Oort cloud being injected periodically into the inner planetary system, some to collide with the Earth. Such collisions can not only cause extinctions of species (as one impact surely did 65 million years ago, killing the dinosaurs), but they could also result in the expulsion of surface material carrying fragments of terrestrially evolved genomes (bacteria and viruses) back into space.

A mechanism thus exists for the genes of evolved Earth life to be transferred to alien planets. A fraction of the Earth debris so expelled survives shock-heating and could be laden with viable microbial ecologies as well as genes of evolved life. Such life-bearing Earth material could reach newly-forming planetary systems in the passing molecular cloud relatively quickly — within a million years of a typical ejection event. A new planetary system will thus come to be infected with terrestrial microbes and terrestrial genes that can contribute, via horizontal gene transfer, to an ongoing process of local biological evolution. In this picture Darwinian evolution is not an Earth-bound process.

Once life has got started and evolved on an alien planet or planets of a new system, the same process of re-infection can be repeated (via comet collisions), transferring genetic material carrying local evolutionary 'experience' via molecular cloud interactions to other nascent planetary systems. Life throughout the galaxy in this picture would constitute a single connected biosphere, in which a pangenome becomes established, ready for re-assembly on any habitable planet by the processes discussed in the last chapter.

How many independent events of origination of life could there have been in the 13.8-billion-year history of the Universe? Can life spread from a single origin in one galaxy to infect the entire Universe? It is in principle possible for bolides or comets to escape from a galaxy of life-bearing particles. The value of the escape speed from a galaxy such as ours cannot be determined precisely, but it must surely exceed the orbital velocity of stars in the outer spiral arms, ~250km/s. Assuming that particles can reach escape speed, say 1,000 km/s (dust could be accelerated by radiation pressure, comets and bolides by gravitational encounters), the distance traversed in the life time of our galaxy in $\sim 10^{10}$ yr is ~10 Mpc. Diffusion of life over much greater distances will be severely limited by "horizon constraints" within an expanding Universe. That is to say, the "edge" of the Universe expands faster than the local speed of the life-bearing dust. Thus life from a single origin in one galaxy might be thought to be contained within a cluster of galaxies defining a pangenome stretching over several Mpc, but not beyond.

In standard Big Bang cosmologies this problem will not be alleviated by going to an earlier epoch of the Universe. In all such cosmological models the first life and its panspermic dispersal can begin only after star formation gets under way and supernovae produce and disperse the heavy elements needed for life. According to the most recent studies this probably happened about 500 million years after the Big Bang, when the Universe was about 5% of its present size, and intergalactic distances were also scaled down by the same factor. Transport times, survival constraints and horizon restrictions although reduced will still inhibit full-scale cosmological panspermia from a single origin.

One way to avoid a horizon-limited spread of life (in a Big Bang cosmology) would be to adopt a non-conformist cosmological model such as that of Gibson and Schild (2009), which involves the creation of Earth-mass clouds 300,000 years after the Big Bang when the totality of material in the Universe changes its state from an ionized condition to a neutral gas. At this point of transition the "cosmological fluid" becomes unstable and breaks up into fragments that Gibson calculates to be of comparable masses to the mass of the Earth. These Earth-mass cloud fragments condense into frozen planets. A fraction of the planets coalesce to form stars, including a proportion of massive stars that evolve on a very short timescale and explode as supernovae. The majority of planets are then polluted with heavy elements from the first generation of supernovae (Gibson *et al.*, 2011).

We have argued that such Earth-mass planets which eventually develop iron cores, overlain with oceans rich in organic materials and weighed down with extensive H-He atmospheres, provide optimal conditions for a first origin of life. The ocean temperature is close to

647 K, the critical temperature of water, and organic synthesis under these ultrahigh pressure, high temperature conditions is greatly speeded up. The volume of a typical such ocean on a primordial planet is $\sim 10^{25}$ cm^3, and with 10^{80} such planets a gigantic cosmological "soup" with a total volume of 10^{105} cm^3 is available for the origin of life. The planetary bodies at this stage are separated by only some tens of AU and with collisional panspermic connections established between them, one could imagine a gigantic inter-connected primordial soup. These optimal conditions would prevail for 10 million years, and cannot be remotely reproduced at any later cosmological epoch.

For judging efficiency in accomplishing an origin of life, the comparison to be made is between the volume of 10^{105} cm^3 for the total set of primordial planets and $\sim 10^{15}$–10^{25} cm^3 for all the hydrothermal vents in the Earth's oceans. A factor of nearly of 10^{90} in probability is thus gained, compared with the factor of 10^{24} we found earlier in going from Earth to all the comets in the galaxy. The creation of life at one locality in this extremely dense, collision-dominated system in the early Universe would lead to the infection of other habitats, and the spread of life across the entire primordial Universe within a few million years. After 10 million years, however, ocean temperatures will drop with the general cooling down of the Universe, and the primordial planets laden with microbes will freeze. The cosmological legacy of life will thereafter be carried within these frozen primordial planets, which Gibson and Schild identify as the baryonic dark matter of the Universe. In the present cosmological epoch such frozen life-bearing planets are located mainly in the halos of galaxies. Mergers and disruptions of such life-laden "giant comets" are associated with every event of star or planet formation occurring in the disc of the galaxy.

To conclude this chapter it should be noted that even the 10^{90} factor gained in going from a primordial soup on Earth to the set of primordial planets in an HGD type cosmology does not entirely overcome the probability hurdle discussed earlier. In Chapter 2 we saw that the probability of $\sim 10^{-180}$ was involved in generating a single ancestral ribozyme comprised of ~300 bases which would still be an underestimate for the progenitor of all the prescriptive information needed for life. According to this analysis an excess of 10^{100} Big Bang universes of the HGD type would be needed to accomplish the certainty of an origin of life occurring in one such universe. We would thus need to contemplate the option of the general class of multiverse models where we just happen by chance to be in the one universe in which the cosmological life-origination event occurred.

An alternative, which in the author's view is philosophically more attractive, involves the class of open cosmological models of which the quasi-steady-state universe (QSS) still remains an option fully consistent with available observations (Hoyle *et al.*, 2000). Somewhere among the infinite amount of material in an open cosmology even a superastronomical improbability will occur, and spread to become an ever-present and integral component of the cosmos. Effectively, then there was no time before which there was no life. Intuitively we may think there must have been, but if we do our instinctive supposition is cultural. A Buddhist for example, might think instinctively that life has always existed for an eternity.

Chapter 4

From Dust to Life

More than 3 decades ago Fred Hoyle and the present writer began to argue, from a purely astronomical standpoint, the case for a widespread occurrence of microbial life in the Universe. On a clear dark night, far from the light of towns and cities, one can see the spectacle of the Milky Way, an edge-on view of our galaxy comprising a few hundred billion stars each more or less similar to the Sun. In addition to stars one can also see dark patches and striations in the Milky Way which are clouds of interstellar dust particles that are so dense as to block out the light of distant stars. Such cosmic dust grains are present in vast quantity everywhere in our galaxy and in external galaxies as well. Serendipitously, it turns out that bacteria on the Earth happen to possess exactly the range of sizes that match these interstellar dust grains.

I shall argue in this chapter that a large fraction of the interstellar dust grains must indeed have been derived from biology. If life is not confined to Earth, and if microorganisms have the ability to infect a multitude of habitable niches throughout the galaxy, such an assertion has a *prima facie* plausibility. The concept of cosmic life is unquestionably a scientific hypothesis that, at the very least, is worthy of investigation and empirical test. The first such test would involve determining the chemical nature of the material present in interstellar clouds. Figure 4.1 shows an image of the Horsehead Nebula which is a dense cloud of interstellar gas and dust. Dust clouds such as this are now known to be rich in organic and inorganic molecules, including water, formaldehyde, and a range of hydrocarbons; and it is in such clouds that new stars and nascent planetary systems are found.

Fig. 4.1. The Horsehead Nebula in the constellation of Orion.

The manner in which starlight is dimmed by the presence of intervening dust clouds was studied by astronomers from as early as the 1930's when astronomical spectroscopy first became available.

In the 1960's one of the outstanding puzzles in astronomy was to understand why the dimming or extinction of starlight over the near infrared to near ultraviolet wavelength range appeared to have a precisely uniform behaviour. The dust grains, thought to be ice grains at the time, had to possess properties that were required to be invariant in regard to size and scattering behaviour everywhere in the galaxy, in whichever direction one looked. This was very difficult to reconcile with any plausible inorganic model of interstellar grains. Interstellar grains made of ice, if they condensed in interstellar gas clouds, would be expected to vary in their average sizes depending on the density of the clouds in which they were formed. The present author's attempts over many years to explain the uniformity of grain properties turned out to be difficult for a wide range of inorganic dust models that were considered.

The points in Fig. 4.2 show the data set that had to be explained in the 1960's in terms of plausible models of the dust. Although ice grains remained the preferred model at the time, astronomical observations in the infrared waveband, which revealed a lack of water-ice absorption, were already beginning to rule out this model by the late 1960's (Wickramasinghe, 1967). With a growing body of evidence favouring an organic dust composition in the late 1970's, Fred Hoyle and the author attempted to solve this problem by invoking grains in the form of hollow desiccated bacteria (as they would be in space) possessing a size distribution and optical properties that matched spore-forming bacteria as were determined in the laboratory (Hoyle and Wickramasinghe, 1982, 1991).

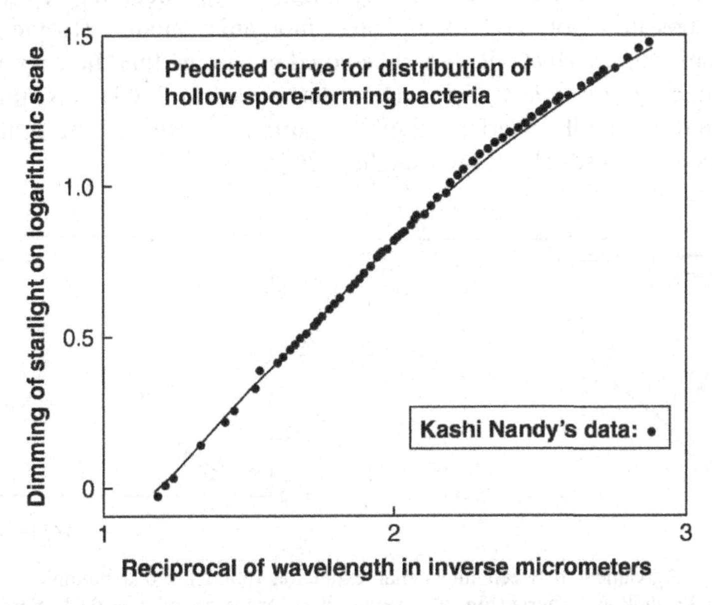

Fig. 4.2. Points represent the visual extinction (Nandy, 1964); the curve is the calculated extinction curve for a size distribution of freeze-dried spore-forming bacteria.

The agreement between the astronomical data points and our bacterial model (Fig. 4.2) was so close that we considered the situation to be pretty close to being decisive. This was particularly compelling because, compared with all earlier inorganic models, we had here a model that was essentially parameter-free. If interstellar dust was bacterial in nature the rest followed, without any further *ad hoc* assumptions being needed.

As often happens, things are never as simple as they appear at first sight. With new observations of the extinction (dimming) of starlight at ultraviolet wavelengths and data in the infrared, refinements to the bacterial model became necessary. We had to introduce a component of dust identified as nanobacteria or viruses, and also their degradation products in the form of aromatic molecules. The former showed up as a continued rise of extinction into the ultraviolet and the latter as a symmetric absorption band centred at a wavelength of 2,175 Å. This composite biological model is displayed in Fig. 4.3, together with the astronomical observations.

Cosmic biology, with no additional *ad hoc* assumptions, matches the entire range of the available data (Figs. 4.2, 4.3) to a degree that was not found possible for any competing inorganic model (Hoyle and Wickramasinghe, 1991). It should be noted in passing that the "hump" in the extinction curve at 2,175 Å (first discovered in 1963) was initially attributed to small spherical graphitic particles arising from coalified bacteria and viruses (Wickramasinghe, 1967).

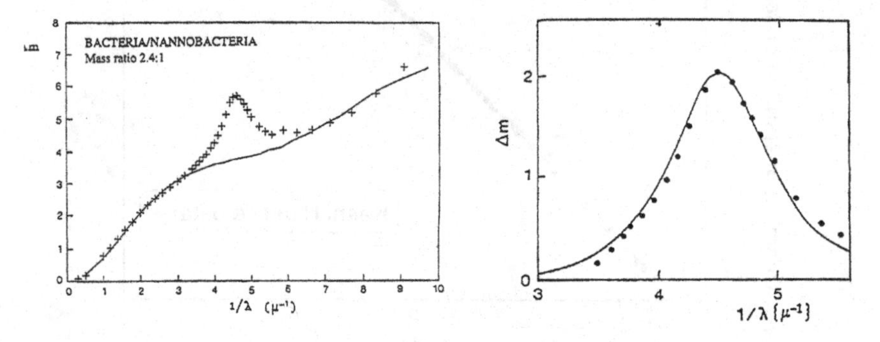

Fig. 4.3. Agreement between interstellar extinction (points) and a biological model comprised of hollow bacteria (Fig. 4.2), nanobacteria or viruses of sizes 0.01 micron and an ensemble of biological aromatic molecules to explain the 2,175 Å hump in the extinction (See Wickramasinghe *et al.*, (2010) for details).

The impressive nature of the correspondences with data shown in Fig. 4.3 was the primary reason for our initial confidence in the biological model of interstellar dust.

A valid scientific theory must be required to make predictions that can be tested and verified, and this was done for many aspects of the model with astoundingly positive results. One prediction was the effect of dust in causing absorption of infrared radiation from distant astronomical sources.

The first such infrared spectrum that was decisively consistent with a prediction from the biological dust model was obtained in 1980–1981 for the Galactic Centre source GC-IRS7 by D.T. Wickramasinghe and D.A. Allen using the Anglo-Australian Telescope (Hoyle and Wickramasinghe, 1992). In view of its historical importance in relation to the evolution of our ideas relating to cosmic biology this comparison is reproduced in Fig. 4.4. The solid curve shows the close correspondence over the waveband 2.8–4 micrometres with the astronomical data obtained by D.T. Wickramasinghe and D.A. Allen. From the measured value of the absorption coefficient of bacteria at 3.4 micrometres and the astronomical data in Fig. 4.4, we can readily infer that about 25–30% of all the interstellar carbon in the line of sight of the galactic centre (the source GC-IRS7) must be in the form of bacterial-type dust grains, or at any rate grains that could not be distinguished spectroscopically from freeze-dried bacteria.

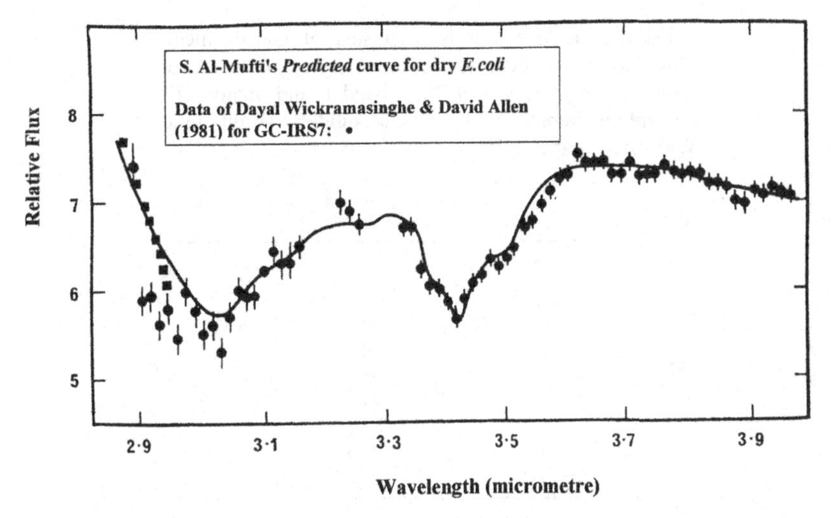

Fig. 4.4. The infrared spectrum of GC-IRS7 showing consistency with desiccated microorganisms over the waveband 2.8–4 micrometres (data points from Allen and Wickramasinghe, 1981).

With improved techniques of observation, the infrared spectra of astronomical sources has been more thoroughly investigated in recent years, for instance by the ESA Infrared Space Observatory and the Spitzer Space Telescope (Smith *et al.*, 2007). Characteristic biological signatures have continued to show up over a wide infrared waveband in a variety of astronomical spectra. Our earlier conclusion that an all-pervasive life-like material is implied in the correspondences shown in Figs. 4.2–4.4 continues to become stronger. Attempts to explain the same data set in terms of abiotic (non-living) processes producing organic molecules, polymers and PAH's remain contrived and unproven.

A set of unidentified infrared emission bands (so-called UIB's) between the wavelengths of 3.3 and 22 μm has been found in almost every dusty region of the galaxy as well as in external galaxies. Recent observations of such UIB's for a large number of galactic and extragalactic sources have been obtained using the Spitzer Space Telescope. These bands occur in the general interstellar medium as well as, more locally, in nebulae associated with the formation of new planetary systems (protoplanetary nebulae, PPN). The comparison of the UIB and PPN wavelengths with spectral features in naturally occurring biological systems is shown in Table 4.1.

Table 4.1. Distribution of two sets of astronomical observations (unidentified infrared bands, UIB's and protoplanetary nebular (PPN) bands) and major IR absorption bands in biological models (Rauf and Wickramasinghe, 2010).

UIBs	PPNe	Algae	Grasses	Bituminous coal	Anthracite coal
3.3	3.3	3.3	–	3.3	3.3
–	3.4	3.4	3.4	3.4	3.4
6.2	6.2	6.0	6.1	6.2	6.2
–	6.9	6.9	6.9	6.9	6.9
–	7.2	7.2	7.2	7.2	7.2
7.7	7.7	–	7.6	–	7.7
–	8.0	8.0	8.0	–	–
8.6	8.6	8.6	–	–	–
11.3	11.3	11.3	11.1	11.5	11.3
–	12.2	12.1	12.05	12.3	12.5
–	13.3	–	–	–	13.4

The origin of the suite of organic molecules that could produce such absorptions is still a subject of debate. Whilst PAH's (polyaromatic hydrocarbons), generally assumed to form inorganically, are currently favoured by astronomers, satisfactory agreement with all available astronomical data has not been possible. The wavelengths of the absorption peaks corresponding to any given PAH (*e.g.* coronene) are known to agree with some of the UIB's, but their precise positions are sensitive to excitation conditions and ionization states. A further constraint on grounds of parsimony is that same set of aromatic molecules that produce the UIB emissions (Table 4.1) must also account for the broad absorption peak at 2,175 Å in the extinction of starlight (see Fig. 4.3). Both sets of observations must arise from the same ensemble of biomolecules.

In addition to the UIB features an emission band at 3.3 micrometres is present, not only in discrete astronomical sources but also as a general diffuse infrared glow of the galaxy. This too has been attributed to PAH's, which in our interpretation must be comprised of biologically derived aromatic molecules. The radiation being emitted in this process must be assumed to be the energy absorbed by the 2,175 Å carrier associated with interstellar extinction. The fact that biological aromatic molecules could act in such a dual role was first noticed over three decades ago (see Hoyle and Wickramasinghe, 1991). Recent laboratory studies by Kani Rauf and the present author have provided further evidence favouring biology and its degradation products in preference to any inorganically produced set of aromatic molecules (Rauf and Wickramasinghe, 2010).

In addition to the absorptions/emissions in the ultraviolet and infrared wavebands the effect of biological aromatic molecules shows up also in the visual spectra of stars. At optical wavelengths, a set of diffuse interstellar absorption bands in stellar spectra, particularly a feature near 4430 Å, poses a continuing enigma to astronomers. These absorption bands have possible explanations on the basis of biologically related molecules such as porphyrins (Hoyle and Wickramasinghe, 1991; Johnson, 1971, 1972). Since their discovery in 1936 the number of these bands, with widths in the range 2–40 Å spanning the waveband from 4,400 Å to 7,000 Å, has risen to well over two dozen. The bands are too wide to be explained in terms of electronic transitions in atoms, ions or small molecules, and to this day their identification and origin remain obscure. The widths and central wavelength placements of the main diffuse interstellar bands are shown in Fig. 4.5.

With a shift of interest from optical astronomy to infrared and ultraviolet observations, the study of these optical bands appears to have been neglected in the past two decades. However, from the considerable data set already available the role of a biological pigment appears to be most likely. In the opinion of the writer the most promising solution is one that was first suggested by F.M. Johnson nearly four decades ago (Johnson, 1972). Even a cursory glance at the spectral data of chlorophylls and metalloporphyrins and related pigments shows that the strongest of the diffuse interstellar bands could indeed arise from such systems — particularly the bands at 4,428 Å, 6,175 Å and 6,614 Å. Johnson has argued that these bands could be due to the specific molecule $MgC_{46}H_{30}N_6$. At a time when no molecule more complex than formaldehyde was known to be present in interstellar space his suggestion was ridiculed and dismissed.

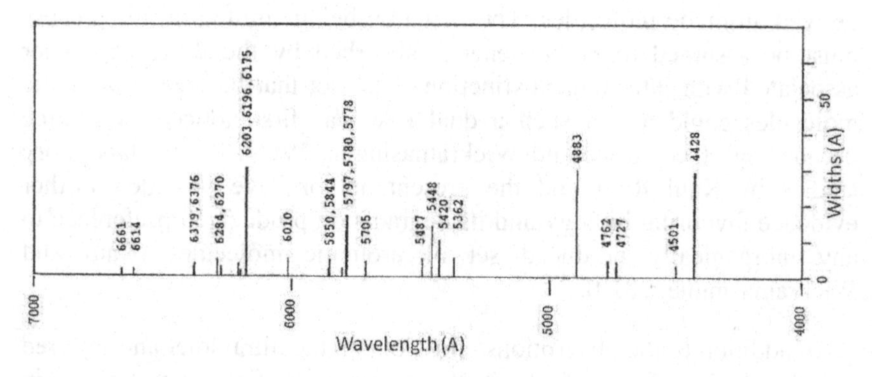

Fig. 4.5. Distribution of main diffuse interstellar absorption bands (see Hoyle and Wickramasinghe, 1991).

Yet another feature of astronomical spectra that appears to be linked to cosmic biology is the so-called extended red emission (ERE). A fluorescence phenomenon over the red waveband 5,000–7,500 Å was first observed in dusty regions of the galaxy by Witt and Schild (1988) and this data has now been considerably expanded. Extended red emission (ERE) is now found in planetary nebulae, ionised hydrogen clouds (HII) regions, dark nebulae and high latitude cirrus clouds in our galaxy, as well as in external galaxies. This phenomenon has a self-consistent explanation on the basis of the fluorescence behaviour of biological chromophores (pigments), *e.g.* chloroplasts and phytochrome. We can show that all these features can be matched to biological pigments. Competing models based on emission by compact inorganic PAH systems are not as satisfactory, as is evident for instance in Fig. 4.6. Hexa-peri-benzocoronene is one of a class of compact polyaromatic hydrocarbons that were discussed in the astronomical literature in this

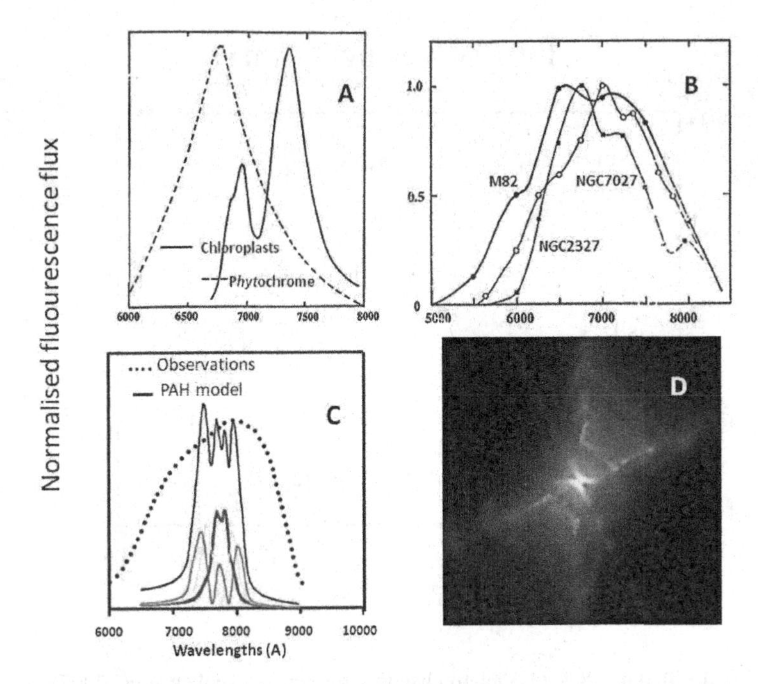

Fig. 4.6. The points in panels B and C show normalised excess flux over scattering continua from data of Furton and Witt (1992) and Perrin *et al.* (1995). Curves in panel C are models of inorganic PAHs; curves in panel A give biological systems — phytochrome and chloroplasts at a temperature of 77 K. Panel D shows a photograph of the Red Rectangle Nebula.

context but their fit is deficient as seen here. Biology still remains the simplest viable explanation of this phenomenon.

To what distances in the Universe does spectroscopic evidence for biology extend? We have seen that our own galaxy is evidently replete with biotic-type material, and that some 25–30% of the carbon in the interstellar clouds may be tied up in this form. Extragalactic sources are also found to have the signatures of living material. These include the extinction properties of dust, which are similar to our own galaxy, the diffuse interstellar bands, the unidentified infrared bands (UIB's) and in some cases the extended red emission (ERE) as well.

Amongst the most distant galaxies displaying aromatic/biomolecular infrared signatures is a high redshift infrared luminous galaxy at redshift $z = 2.69$, the spectrum of which is shown in Fig. 4.7 (Tepletz *et al.*,

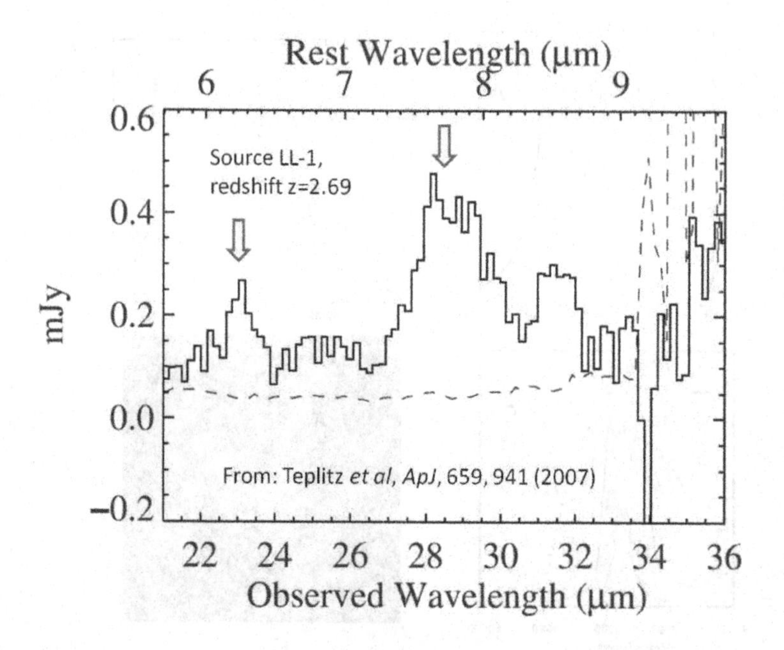

Fig. 4.7. Redshifted 6.2, 8.7, 11.3 micron bands in the source (Tepletz *et al.*, 2007).

2007). This galaxy emitted its light when the Universe was at the tender age of 2.5 billion years according to standard Big Bang cosmology.

The ultraviolet hump in interstellar dust absorption at 2,175 Å, that can also be assigned to biomolecules, shows up in observations of the most distant galaxies upto redshifts of 2.45 (distances of 11 billion light years). This demonstrates clearly that biologically related materials were formed within 2.5 billion years of the presumed Big Bang origin of the Universe (Ellasdottir *et al.*, 2009; Motta *et al.*, 2002; Noterdaeme *et al.*, 2009).

To conclude this chapter we note that in the five decades of the author's research into the nature of interstellar dust grains there have been many unexpected shifts of paradigm. The ice-grain theory proposed by H.C. van de Hulst seemed set in stone when the author first began his studies in 1961 (van de Hulst, 1949), but its overthrow did eventually occur. The introduction of the graphite particle theory, that was initially vigorously opposed, gained ground over a timescale of a decade (Hoyle and Wickramasinghe, 1962). But soon afterwards, with the emergence of more refined astronomical observations and calculations, Hoyle and the present author felt obliged to abandon our simpler graphite model in favour of polymeric dust grains and aromatic molecules as we have just discussed (Wickramasinghe, 1974; Hoyle and Wickramasinghe, 1977).

Mineral silicates as a component of the grains came rapidly into vogue from 1969 onwards. However, pinning down a particular mineral silicate that matched infrared data proved so difficult that astronomers felt obliged to postulate the existence of an "astronomical silicate" — thus inverting the problem and defining a hypothetical absorbing material to match astronomical spectra over the 8–12 μm wave band. Although the presence of silicate particles in interstellar space cannot be denied (we are standing on a silicate planet!), the overwhelming dominance of organics and organic polymers over the wavebands 3–4 μm, 8–14 μm and 18–22 μm has been reaffirmed over many years (Hoyle and Wickramasinghe, 1991). The current trend to consider mixtures of graphite particles and silicate grains as an explanation of the data in Figs. 4.2 and 4.3 is unsatisfactory in the author's view (see Draine, 2003). We ourselves invented this model and this procedure in 1969 (Hoyle and Wickramasinghe, 1969), but later abandoned it in favour of the biological model for the reasons discussed in this chapter. The best graphite-silicate model requires finely tuned size distributions for both components, and then the problem is to understand why these quantities are so precisely invariant over much of the galaxy. A biological model explains such invariance naturally and elegantly.

The universe is surely a far stranger place than we could ever imagine and many more surprises must lie in store. The 1970's, 1980's and 1990's witnessed the unravelling of a veritable Pandora's box of complex organics in every nook and cranny of the cosmos. With a large fraction of all the carbon in the universe locked up as organic, life-like material, the connection with biology seemed inevitable. Two choices remain. Are we to suppose that we are witnessing an incredibly difficult, near impossible, progression from non-life to life occurring *everywhere* in space? Or could we be seeing evidence for the omnipresence of cosmic life and the inevitability of panspermia? The present author opted for the latter — life being a cosmic phenomenon — and evidence to support this position continues to grow.

The biological theory of grains has the distinct merit of serving as a unifying hypothesis for a vast range of observations. Alternative non-biological explanations appear highly contrived, demanding fine-tuning of large numbers of free parameters. The situation is strikingly reminiscent of the Ptolemaic model of the Solar System half a millennium ago — a new epicycle being required for every new observation, if an outmoded paradigm is to be retained.

Many astrobiologists have unwisely chosen to disconnect astronomical phenomena from biology. The trend is to interpret a vast body of evidence for biochemicals in interstellar space as evidence of a *hypothetical* prebiotic chemistry operating on an astronomical scale. Such a presumption has no factual basis whatsoever. It is far more probable that the origin of life was a unique event that happened only once in the history of the Universe. The degradation of living cells is a well-understood process: the transformation of organic life on the Earth through a series of steps leading eventually to anthracite and coal is also well documented. The astronomical data discussed in this chapter — the 2,175 Å absorption, the unidentified infrared and visual bands and ERE — all have a far better chance of being correctly explained as evidence of biology rather than prebiology.

Chapter 5

Comets

Some four and a half billion years ago the Earth was in its final stages of assembly from rocky debris that occupied the inner regions of our solar nebula. The rocky outer crust, the mantle below and the metallic core were all in place, but the final touches to a structure that was to become our home for life were still to be finished. At this time the Earth probably turned on its axis somewhat faster than it does now — the day was thus shorter than 24 hours — and the sky was frequently scribed with the majestic trails of comets. The outer icy planets were still in the process of being assembled from their comet-type progenitors. Many a comet would have crashed onto the rocky surface of the young planet and in the process brought a rich bounty of volatile materials such as water and carbon dioxide that went to form the Earth's primitive oceans and atmosphere. Cometary and asteroidal bodies cratered and scarred the surface as they struck, but they also eventually transformed a barren lunar-type landscape of the primitive Earth into a planet with water and blue skies, creating at last a congenial home for life.

The impact history of the Earth is schematically depicted in Fig. 5.1. The asymptotic rise of impact rate near the origin of axes $t = 4.5$ billion years ago represents the final stages of the accumulation of the planet's outer layers. The spike in the rate of impacts at about 4 billion years ago corresponds to the Moon-forming event when an exceptionally large impactor is thought to have blasted off material that recondensed to form a satellite.

We mentioned earlier that by what seemed a remarkable act of providence the very oldest evidence of terrestrial life coincides almost exactly with the moment when cometary bombardment had dwindled to an insubstantial trickle. This was a time between 3.83 and 4 billion years

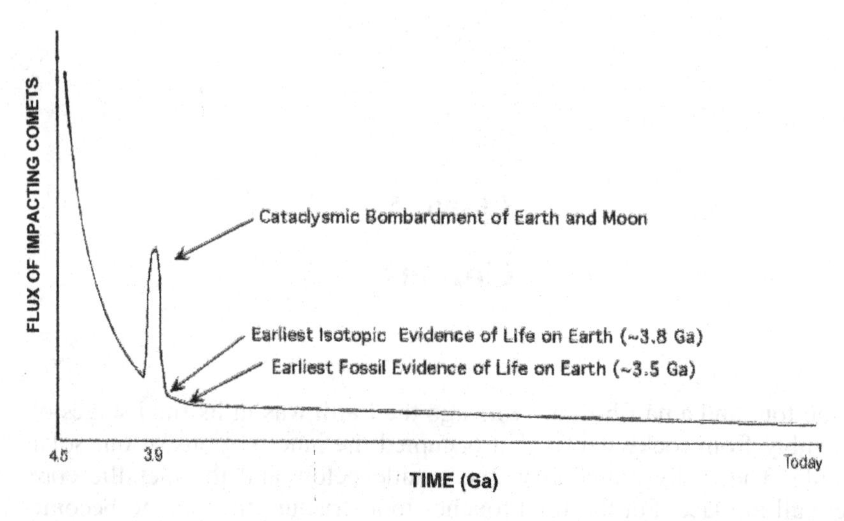

Fig. 5.1. Schematic plot of the frequency of cratering impacts on the Earth.

ago when the last phase of the accumulation of the planet from cometary material was complete. The only rational way to interpret these facts is to assume that life was added to Earth by impacting comets, and that cometary life took root on our planet as soon as the conditions on the surface became congenial. There is no evidence for any traces of a geological epoch in which the so-called primordial soup existed. Nor is there evidence of any prebiotic chemical evolution on Earth that predated the appearance of the oldest life.

In this picture it would be natural to regard comets as the most likely bringers of life to the Earth, and more generally as the carriers and distributors of life and of life's genetic heritage throughout the cosmos. Fred Hoyle and the present author arrived at this position in 1981 when evidence for bacteria or bacterial type dust in interstellar space seems to have been established beyond reasonable doubt as we saw in Chapter 4.

Having thus challenged the time-hallowed paradigm of the primordial soup, turning to comets for the source of life was in a sense a double transgression in the eyes of our critics. A connection between comets and human affairs had been deeply ingrained in diverse cultures worldwide and an ancient fear of comets appears to be almost universal. Attempts by Hoyle and the present author to link comets to the origin of earthly life were too much like a revival of an ancient superstition, and so we were met with general disapproval at this stage.

Ancient ideas on comets fall into two categories. They were either thought of as astronomical objects on a par with the planets, or they were regarded as meteorological phenomena — incandescent vapours that existed only in the atmosphere. In either case the focus was not on the primary phenomenon itself but on the effects it generated. Aristotle who argued that comets are fiery meteors probably came nearest to the truth. He further claimed that a cometary apparition ushered in a range of terrestrial effects — stones falling from the sky, storms, droughts, tidal waves and even earthquakes. Once again he seems to have been nearer to possible truths about such matters than later thinkers had been. Considering his errors and misjudgements in relation to both biology and cosmology, to which we have referred earlier, this was indeed a welcome change.

The ancient belief that comets heralded unpleasant effects led to these celestial objects being viewed as omens. Thus Shakespeare wrote:

> "When beggars die there are no comets seen;
> The heavens themselves blaze forth the death of princes."

Such early ideas about comets must have been a complex mixture of fact and wrong inference from observing certain coincidences, some of which were perhaps meaningful, others not. Poetic license may have played a role, as well as superstition and fear. This would have been especially so as the nature of the cometary phenomenon remained well beyond comprehension until relatively recent times.

The periodicity of cometary appearances was first noted by the English astronomer Edmund Halley (1656–1742). He observed close similarities between the orbital paths of three comets that appeared in 1531, 1607 and 1682 and inferred that it was the same comet making repeated circuits round the Sun. By applying Newton's theory of gravitation, the French astronomer A.-C. Clairaunt (1713–65) was able to predict that the same comet would reappear in 1759, and it surely did. This event of verification provided a convincing vindication of Newton's theory of gravitation. Studies of Newton and Halley left little doubt that cometary bodies are substantially smaller than planets and that they pursued eccentric elliptical orbits around the Sun, this behaviour being a consequence of the inverse square law of gravitation.

Despite the sophistication of seventeenth-century Newtonian mechanics, and the general emergence of a mechanistic, Cartesian world view, the need for cometary divination and superstition related to comets continued. Newton himself noted that comets move in chaotic elliptic orbits which are randomly oriented with respect to the plane of orbits of the planets (the ecliptic plane). He was therefore acutely aware of the danger of comets colliding with planets with consequent catastrophic effects. However, he took solace in the belief that Providence or a benevolent God would have arranged matters in such a way that we would be spared from devastating collisions except perhaps on the rarest of occasions. The latter too he thought was important in a positive sense of providing a periodic refueling of the Earth and planets.

In the late nineteenth and early twentieth century the studies of comets were confined to amateurs and small groups of committed professional astronomers and physicists. Cometary science was regarded as largely irrelevant to the understanding of the nature of stars, planets and the larger-scale structure of the Universe. It thus remained very much on the fringes of astronomy and astrophysics. The dangers to terrestrial life, that were dimly apprehended by Newton and expressed in his correspondence, were largely forgotten, or more likely ignored. Comets were, on account of their relatively small masses, considered as inconsequential cosmic objects. Whilst an individual comet measuring 10 kilometres across might be considered small in relation to other astronomical bodies, a hundred billion of them, which exist in the outermost reaches of the solar system, are not inconsequential and by no means negligible in their effects. Studies over the past decades have elevated the status of comets to an important status in astronomy.

Comets move in elliptical orbits around the Sun and have orbital periods that range from a few years to hundreds of thousands of years. Short-period comets originate in a flat disc known as the Kuiper belt which lies beyond the orbit of Neptune, whereas the long-period comets appear to originate in a spherical shell of icy bodies in the outer solar system located at radial distances between 50,000 and a few hundred thousand astronomical units (1 astronomical unit = the mean distance between the Sun and the Earth). This is depicted schematically in Fig. 5.2.

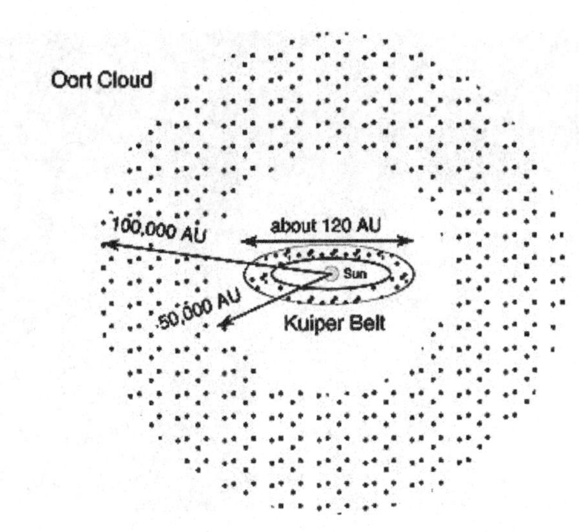

Fig. 5.2. Schematic distribution of long and short-period comets in the solar system.

Comets from the Oort cloud plunge towards the Sun at an average rate of a few per year due to random gravitational perturbations caused by the massive outer planets or by passing stars. This rate of ingress is greatly enhanced whenever the solar system passes close by a massive molecular cloud. Such occurrences are estimated to take place on the average of once over 40–50 million years. Whenever this happens the rate of comet impacts onto the inner planets, including Earth, is greatly increased, and it has been argued that such impacts could lead to extinctions of species as well as splash back of surface dust and rocks into space (Wickramasinghe *et al.*, 2010). We discuss this process continuing to recent times in a later chapter.

As a comet approaches the Sun in its orbit, solar heat begins to vaporise the surface material. The liberated gases surrounding the nucleus soon begin to glow, producing a fuzzy coma typically extending out to half a million kilometres, a dimension that compares with the size of the Sun. The ultraviolet light from the Sun and the outward moving gas from the solar surface then blow the luminous gas and dust in the comet's coma into a tail that can itself extend to lengths of 10 to 100 million kilometres. The tail of a comet often splits into two — one wispy thin tail made of gas, and a broader, gracefully curving fan-shaped tail made of dust, the dust tail often fanning out over an angle of

Fig. 5.3. Comet Hale–Bopp reached perihelion on 1997 April 1 (period 2,520 years, aphelion 370 astronomical units).

10–20 degrees across the sky. The result, particularly for a large comet, is one of the most spectacular sights visible in the night sky.

Once we conclude that interstellar dust involved biologically generated material as we discussed in Chapter 4, the route to a connection of comets with life seemed more or less well-defined. In the 1970's comets were thought of as uninteresting lumps of dirty ice — a model advocated by Fred Whipple, and one that had come to be regarded as gospel truth in the scientific community for over two decades. All that was soon to change with the arrival of new observational evidence, but the final abandonment of the dirty-ice Whipple paradigm was a much harder and slower process.

Whipple's dirty ice conglomerate or dirty iceberg model came into vogue after the discovery, in cometary comas and tails, of a number of emission bands corresponding to many atomic species and simple molecules such as water, carbon monoxide and various radicals. It seemed reasonable then to suppose that a mixture of water, methane and ammonia in frozen form served as a source of the parent molecules in the comet's nucleus. Cometary tails had been shown to have the spectral characteristics of sunlight scattered from small particles about a micron in size. These solid particles were thought to be mixed in with the more volatile material that makes up the bulk of the nucleus of a comet.

As the volatile gaseous material is expelled from the nucleus, it carries the dust component along with it. In the 1970's the dust in the tails of comets was thought to be made up of mineral silicates. A crude representation of the Whipple dirty iceberg comet was a 10 kilometre wide chunk of frozen water, methane and ammonia with an admixture of silicate dust and other molecules sitting lightly within microscopic cavities of an icy matrix.

As we progressed through the early 1970's, radio astronomers were discovering molecules of formaldehyde (H_2CO), methyl cyanide (CH_3CN) and hydrogen cyanide (HCN) in the gases emanating from comets. The new data did not fit neatly into the reigning cometary paradigm and cast doubt on the entire conventional model. At this time the present author became acquainted with Professor V. Vanysek of Charles University in Prague who had been studying the spectra of comets and had himself begun to feel uncomfortable about the Whipple dirty iceberg model. Vanysek and the author thereupon proposed a radically new model for dust in comets. We suggested that the bulk of the cometary nucleus was made of refractory organic material, with organic polymers breaking loose at the surface and further fragmenting into smaller molecular units as the comet came close enough to the Sun (Vanysek and Wickramasinghe, 1975).

After a lapse of several years Fred Hoyle and the author began to contemplate much deeper issues that related to comets. What if the comet dust was not just composed of non-biological organic polymers, but intact bacteria and their degradation products? This suggested a connection between interstellar dust, which we had argued was biogenically derived, and the dust seen in the tails of comets. The status of comets would immediately be raised from the insignificant objects they had been once thought to be to the most interesting and important class of objects in the entire universe. These dust grains possessing biological characteristics may themselves have been derived from the frozen primordial planets dating back to a few million years after the Big Bang origin of the Universe, as discussed in Chapter 3; or they may be an ever-present attribute in a quasi-steady-state type universe with an open timescale.

We now proceed to develop the concept of comets serving as microbial incubators in greater detail. If the original comet reservoir, the Oort cloud of comets in our solar system, had incorporated even a trillionth in mass of included dust grains that were in the form of viable bacteria, there would be hundreds of viable bacterial cells per comet at the outset.

Newly condensed comets in our solar system would have possessed liquefied interiors kept warm by radioactive decays lasting typically for the order of millions of years in the manner we discussed in an earlier chapter. We could thus think of each comet as a gigantic culture medium containing at the outset a few viable cells in the midst of water, organic nutrients and all the necessary inorganic salts needed for rapid microbial growth. The volume of this culture medium would be vastly bigger that anything that could be imagined in a biochemist's laboratory, bigger than the Empire State Building, bigger than the largest orbiting space station that is planned for the foreseeable future. For as long as an interior region of a comet remains melted, these biologically favourable conditions within an individual comet seem inescapable.

The length of time for which the interior liquid state is maintained within a comet depends on the radioactive sources that are available and the overall size of the comet. An outer shell of thickness about a kilometre in a 10 km radius comet could provide enough insulation to preserve a warm watery interior for a long enough time for a replicating bacterium to be able to swamp and colonise the bulk of a comet's interior volume. If one imagines a single viable anaerobic bacterium placed within the liquid interior of a comet, it would divide to yield two offspring in two to three hours. These then divide sequentially every two hours, two becoming four, four becoming eight, eight becoming sixteen and so on. With 40 doublings and with continued access to nutrient, the culture expands to the size of a marble in sugar cube in four days. With eighty doublings the culture would grow to the size of a village pond in eight days, and with only 120 doublings an entire cometary core would be converted into biological material in mere 12 days. The precise timescales in this calculation are underestimates of course because we assumed instant access to nutrients and energy at all times. A more realistic timescale of cometary transformation into biomaterial may well be in the order of hundreds of years, but this is well under the timescales that are available for the preservation of liquid interiors in comets. And so it follows that the 100 billion comets in the early history of the solar system would have provided the most powerful setting for amplifying a population of primordial cosmic bacteria — for regenerating and perpetuating the legacy of cosmic life. With each one of 100 billion Sun-like stars in our galaxy providing similar conditions, the scheme for amplifying and maintaining cosmic biology on a cosmological scale is impressively powerful.

How, one might ask, could this idea be tested? The first opportunity arose in the run-up to the 1986 return of Halley's Comet. It would seem entirely appropriate that an important paradigm shift relating to the composition and cosmic role of comets came to be associated with this truly historic comet. As mentioned earlier this was the comet that Edmund Halley had studied in 1531, 1607 and 1682 that led to the realisation that comets pursued elliptical orbits around the Sun. It was also the comet that provided confirmation of Newton's theory of gravitation and effectively clinched the Copernican revolution in science. Halley's Comet has an average orbital period of about 76 years, and its last perihelion passage or closest approach to the Sun took place in February 1986.

Ahead of the European spacecraft Giotto reaching Halley's Comet, Hoyle and the author published a paper entitled "Some Predictions of Comet Halley" early in 1986 (Hoyle and Wickramasinghe, 1986). Here we made predictions of what the cameras aboard Giotto would see at closest approach to the nucleus of the comet. We argued that the surface would look dark, roughened and generally coal-like, the result of the prolonged weathering and effective coalification of biological polymers. That is precisely what was found, a surface that was described as "black as coal" by the project scientists— albedo less than 0.01 percent. The architects of the Giotto Comet Halley project paid a heavy price for their blind faith in a wrong theory. They expected to find a surface as bright as a snowfield and accordingly trained their cameras to focus on the points of greatest intensity. The dark surface of the comet therefore caused a problem in the pictures that were initially beamed, leading to the disappointment of many millions who watched the Giotto–Halley encounter on their TV sets. Later processing of the images revealed a giant peanut-shaped image of the nucleus of Halley's comet measuring some 16 × 8 × 7 kilometres — the first comet nucleus ever to be seen and imaged at close quarters (Fig. 5.4).

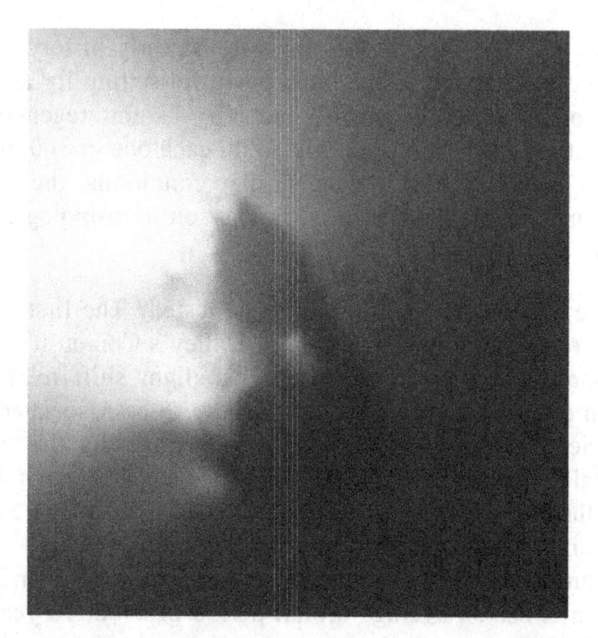

Fig. 5.4. Early image of Comet Halley (1986).

Although the millions of dollars spent on the Giotto mission did eventually lead to a wide range of important results, the first strong support for the cometary life hypothesis followed from ground-based observations made by Dayal Wickramasinghe (the author's brother) and David Allen at the Anglo-Australian Telescope on 31 March 1986 (Wickramasinghe and Allen, 1986). Their data on the infrared spectrum of the dust from Comet Halley matched perfectly the laboratory spectrum of heated bacterial dust as seen in Fig. 5.5.

Comet Halley was expelling bacterial-type dust at the rate of a million or more tonnes per day when this observation was made. And this is what Comet Halley kept on doing for almost as long as it remained within observational range. An independent analysis of dust impacting on mass spectrometers aboard the spacecraft Giotto also showed a complex organic composition of comet dust, a composition that was fully consistent with a biological model. Broadly similar conclusions have been shown to be valid for other comets as well, in particular Comet Hyakutake and Comet Hale–Bopp.

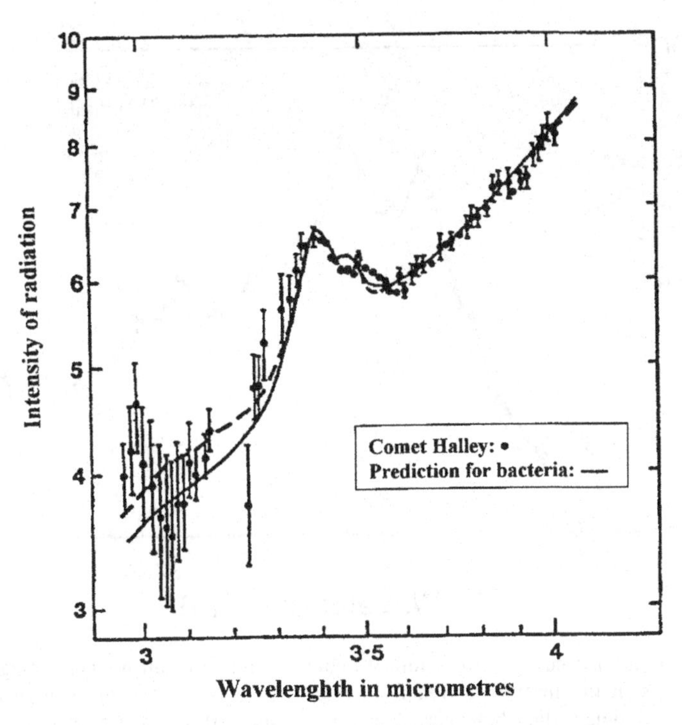

Fig. 5.5. Emission by the dust coma of Comet Halley observed by D.T. Wickramasinghe and D.A. Allen on March 31, 1986 (points) compared with bacterial models. Calculated curves are from Hoyle and Wickramasinghe (1991).

As with the introduction of every new observing technique, the use of ISO (Infrared Space Observatory), launched by ESA on 17 November 1995, provided new opportunities for testing cometary theories. Spectral features near 19, 24, 28, and 34 μm observed by ISO in Comet Hale–Bopp were attributed to hydrated silicates, but the uniqueness of some of these assignments is still in doubt. However, even on the basis of a silicate identification of the principal infrared bands, such material could make up only some few percent of the mass of the dust. This appears to be the case for the infrared flux curve of Comet Hale–Bopp, obtained by Crovisier *et al.* (1997) when the comet was at a heliocentric distance of 2.9 AU. The jagged data curve in Fig. 5.6 may at first sight imply an overwhelming dominance of olivine grains but detailed modelling showed otherwise.

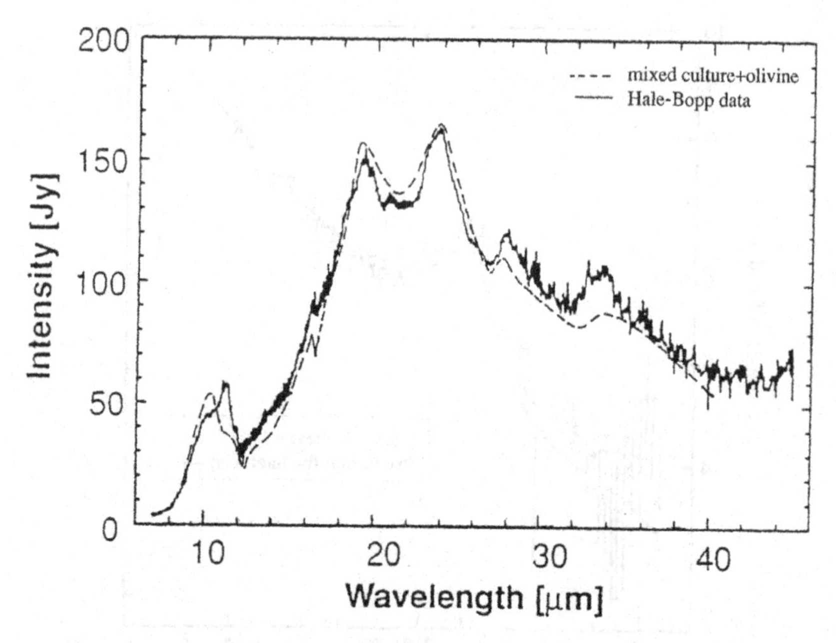

Fig. 5.6. The dashed curve is for a mixed culture of microorganisms containing about 20% by mass in the form of diatoms. Olivine dust, which has a much higher mass absorption coefficient than biomaterial, makes up only 10% of the total mass in this model (Wickramasinghe *et al.*, 2010).

Since 1986 many space probes have gone to other comets and so far the data we have are all generally consistent with what was found from Comet Halley. All have exceedingly low albedos and dust with very similar spectroscopic properties. Figure 5.7 shows a montage of images of comets at various perihelion distances q compared with Comet Halley.

The composition and structure of the crust and sub-crustal layers of a comet came sharply into focus on 4 July 2005. NASA's Deep Impact mission dispatched a 370 kg projectile at a speed of ~10 km/s to crash head-on into Comet Tempel 1. Large quantities of gas and grains were expelled to form an extended plume and coma. Spectra of the coma in the near IR waveband 4 minutes after the impact showed a sharp increase of emission by CH bonds in organic dust over the waveband 3.3–3.5 μm relative to an approximately constant water signal as shown in Fig. 5.8.

The excess radiation over this waveband in the post-impact plume that cannot be modelled by inorganic coma gases is best explained on the basis of degraded biologic-type organic material (Lisse *et al.*, 2006).

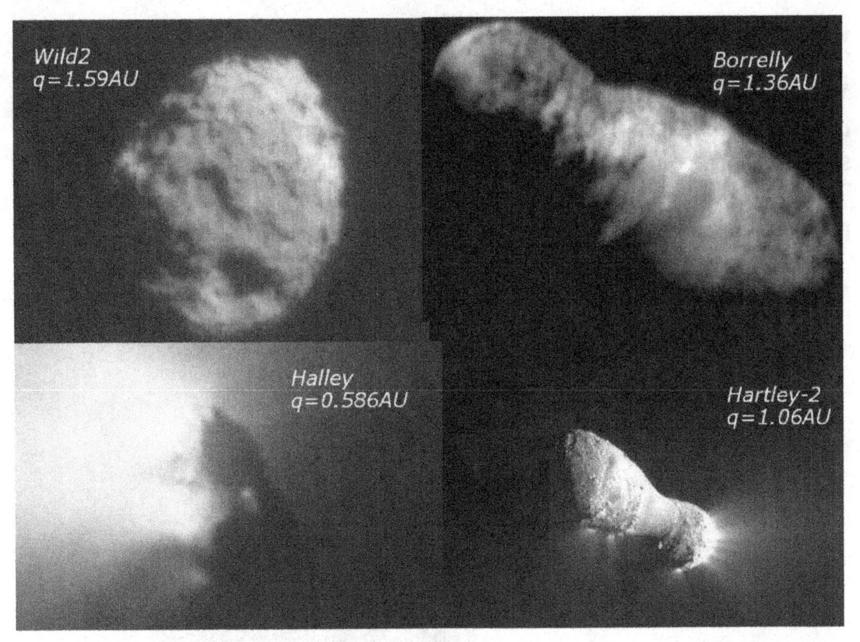

Fig. 5.7. Montage of images of several comets at various perihelion distances q.

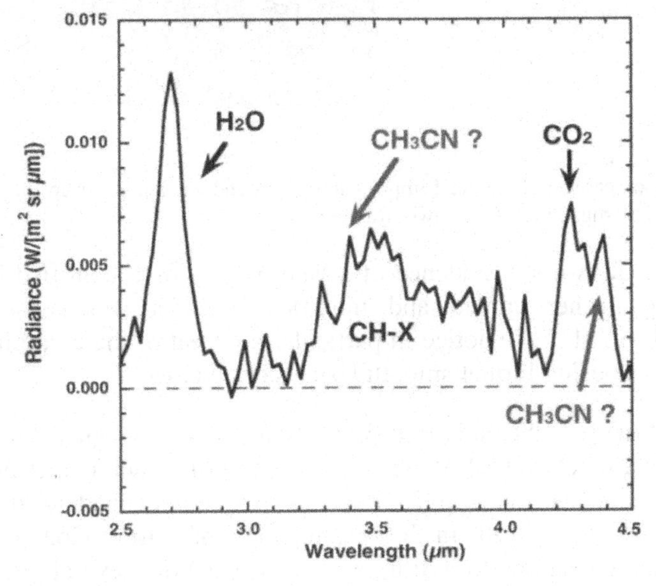

Fig. 5.8. Spectrum of coma of Comet Tempel 1 four minutes after impact (A'Hearn *et al.*, 2005).

Evidence of Cometary Lakes in Comet Tempel 1

Whenever overlying layers of insulating crust of comets become dislodged or disrupted, due to internal activity or meteorite impacts, smooth surfaces of internal lakes can be exposed. Areas of Comet Tempel 1 photographed by cameras on the Deep Impact spacecraft reveal precisely such characteristics (Wickramasinghe *et al.*, 2010).

Comet Tempel-1
2005
2005
2011
The major lake-plateau identified by Deep Impact in 2005 (upper image) appears eroded by up to 30 metres on the right, of 2011 (lower)

Fig. 5.9. Comparison of Comet Tempel 1 terrain around a crater — Deep Impact, 2005 (upper) and Stardust, Feb 2011 (lower frames).

In Fig. 5.9 we see evidence of a lake plateau first identified in 2005 becoming further eroded and modified when it was subsequently observed in 2011. We notice in particular a retreat of the escarpment on the right, exposing further smooth lake surface below.

In February 1999 the space probe Stardust set out on a seven-year voyage to Comet Wild 2. Its mission was to bring back comet dust that impacted and became trapped in blocks of aerogel, and to study the fragments that survived. In 2006 samples of dust from Comet Wild 2 were returned safely to Earth, and distributed to several groups of scientists around the world for analysis. As expected the high velocities of impact onto the collecting aerogel cells left little evidence of any original organic grains or putative cells — only trails of molecular debris. Whilst no living cells were recovered, complex organic molecules were found in abundance in the debris trails, including an amino acid glycine; and all this was consistent with the break-up of biological material. The biological explanation for the genesis of this material appears more plausible, according to the thesis of the present book, than the claim that the organics may represent products of radiation processing of simpler molecules. In addition to organics, the collected material also contained mineral particles, including the mineral cubite. Since cubite can only be formed in the presence of liquid water, this discovery provides clear evidence that liquid water existed in primordial comets to play a crucial role in the replication of cometary bacteria as we discussed in this chapter.

Although microorganisms in comets are expected to be in a frozen dormant state at very low temperatures (less than 50 degrees Kelvin) a sporadic resumption of metabolism will occur if subsurface melting can take place. Such melting could take place when comets come close enough to the Sun at perihelion, or on occasions when impacts of smaller bodies transfer kinetic energy that can be converted to heat. Comet Hale–Bopp showed sporadic activity when it was outside the orbit of Jupiter in the cold depths of space. We can argue that this evidence points to the resumption of bacterial activity (Wickramasinghe, Hoyle and Lloyd, 1996). In our model metabolism builds up subsurface gas pressures of thousands of atmospheres, which is enough to cause cracks in the frozen surface crust releasing gas and dust.

Direct photographic evidence of water jets from a comet was obtained in 2011 when the Stardust spacecraft (that rendezvoused with Comet Wild 2 in 2004) came close to the Comet Tempel 1. Cameras aboard the spacecraft captured images of jets that emanated from surface fissures (Fig. 5.10). The process suggested here may be more immediately familiar if one thinks of the “popping” or bulging out of a can of food in which bacterial action is taking place — so called “spoilt food”. Pressures approaching 10^3 atmospheres easily occur in such cases, similar to the pressure in comets. An even more dramatic example is seen in the explosion of a bottle of wine in which excessive fermentation has occurred. A wine bottle, once ruptured, stays ruptured of course; whereas the surface layers of a comet re-freeze back to their initial tensile strength, providing for a repetition of the phenomenon many

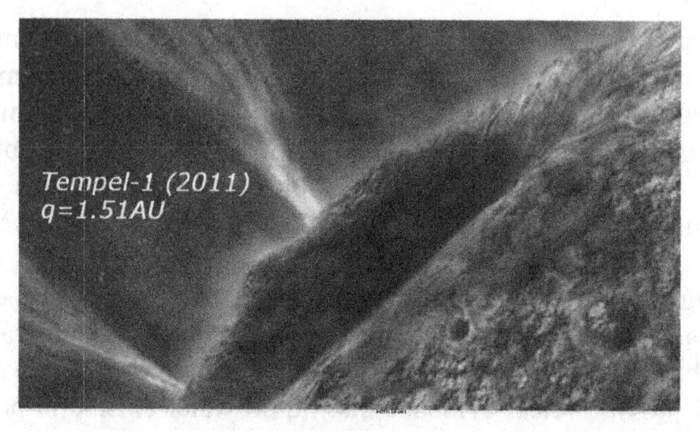

Fig. 5.10. Comet Tempel 1 photographed by cameras aboard Stardust spacecraft in Feb 2011 showing jets of water emerging from fissures.

times. Although in different settings, the processes are similar. We conclude that cometary dust particles, just like the interstellar dust, are indistinguishable from bacteria, a fraction of which are in a frozen dormant state.

The European Space Agency's Rosetta mission to Comet 67P/Churyumov–Gerasimenko was perhaps the most ambitious space mission to any comet. The rendezvous of the Rosetta spacecraft with the comet on August 7, 2014 revealed a dark, rough dumbbell-shaped object 4 km in length, rotating with a period of 12.7 hours. Earlier in June 2014, when the average surface temperature would have been 130 K, the comet was reported to have been fizzing out steam at the rate of about 300 ml every second, thus implying the presence of super-heated interior domains. The situation would be similar to what we discussed in relation to Comet Tempel 1 and may well be an indication of the existence of thermophilic microorganisms.

Although the Rosetta mission, and its lander Philae which is to be set down in November 2014, did not carry any life detection experiment, there is little doubt that indirect evidence for biology will be forthcoming. Figure 5.11 shows a combination of dark roughened regions with smooth areas that have the appearance of exposed subsurface lakes. The comet's morphology and composition as well as high level of chemical activity, if confirmed, could all point to biology.

Fig. 5.11. Image of surface of Comet 67P/Churyumov–Gerasimenko obtained by Rosetta cameras on 7 August 2014 (Courtesy of ESA).

Chapter 6

Cosmic Viruses in Our Genes

We have seen that the pattern of evolution of life on the Earth is characterised by long periods of very slow change, punctuated by episodes of rapid and abrupt change. The latter is a combination of huge surges of innovation involving the introduction of new suites of genes, the emergence of new species as well as the termination or extinction of others. The long periods of slow evolution or stasis is easily understood in terms of Darwinian evolution in a "closed box" context, whereas the sharp changes are more reasonably interpreted as pointing to the ingress of new genes — viruses and bacteria — from space. The latter is important in order to explain the steady evolutionary progress over time, including the development of complexity and the diversification of lifeforms on our planet. Viruses and viral genes are likely to play the most important role in this process. They act in their well-attested role as mobile genetic elements, stretches of coded RNA or DNA that can move between living cells, on Earth, and indeed everywhere in the cosmos.

A curious property of our DNA that has only recently come to light is that over 90 percent is normally inert — that is to say, it is not used for making proteins but is merely copied repeatedly from cell to cell, from generation to generation. In certain relatively rare diseases viral particles are seen to emerge from this normally inert DNA, thus suggesting that the whole of our DNA might indeed be derived from viruses. The point of view that we shall develop here is that DNA, the blueprint of all life, was assembled from genetic fragments that actually came from the wider cosmos. Darwinian evolution is then to be regarded as a process that is not confined to the biosphere of our planet.

A typical virus is a tiny particle measuring about one ten-thousandth or less than the size of a pinhead. It is made up of nucleic acid, either RNA or DNA, and is surrounded in most but not all cases by a double-layered shell of protein aceous material. The external shapes of viruses are regular geometrical figures, a common form being the icosahedron, a solid which has 20 triangular faces. Spikes protrude from the corners of this solid which are extensions of the protein shell that actually help the virus to recognise suitable hosts and eventually penetrate a host cell. There is an intimate cell–virus relationship, a relationship that is almost conspiratorial.

The virus attaches itself to specific sites on the surface of a host cell. Then it is quickly engulfed by the cell's outer membrane and is effectively sucked into the interior of the cell. Next, the host cell proceeds to strip the virus of its outer protein coat, and thereafter it takes its instructions from the invading virus. The instruction is effectively "stop what you are doing and produce more viruses like me!" This instruction is instantly obeyed by the cell. Finally, the newly-formed viral particles burst forth from the cell wall, almost invariably destroying the cell in the process and causing other host cells to be infected by the viral particles so released.

Viruses are in general quite selective about the kind of cells that they attack. A particular virus — *e.g.* the influenza virus — selects a species or a narrow range of species, and also a subset of cells within a species. This property is widely cited as an argument against an extraterrestrial origin of viruses. One could justifiably ask the question: how could a virus that evolved in some place other than the Earth replicate in terrestrial host cells? In other words, how could a virus arriving at the Earth know ahead of its coming here the nature of the cells which it is going to find? The answer is simple. It is true that the incoming virus cannot know in advance what host cell it is going to encounter. But we, the host cells could recognise the virus, if our own genes contain viruses of a similar kind. According to this point of view we must have had a long and continuous interaction with viruses in our evolutionary history stretching back over billions of years. Thus a virus such as the influenza virus (or any other virus) that comes in today simply seeks out those species in whose ancestral lines certain aspects of the same virus were incorporated perhaps millions of years ago.

Within the context of an Earth-bound theory of life, this apparent conspiracy on the part of cells of higher life forms to admit viruses is baffling. If there was no favourable positive aspect to the process of viral infection, it is surprising that evolution could not have conferred total protection or immunity from viral attack. In the many steps of the virus–cell interaction a blockage of access at any one step could have been easily arranged. The logical possibility of preventing the overwriting of the cell's genetic programme must surely exist, for the much larger information content of the host cell could very easily contrive to swamp the trivial information content of an invading virus. If viruses had no positive role to play it is hard to imagine that such a defence was not developed over long evolutionary timescales in creatures that are as highly evolved as we seem to be. The reason that this has not happened must be that the entry of viruses into our genomes was a prime requirement for evolution, and was thus to be encouraged rather than prevented.

This point of view is supported by considering the specificity of viruses to particular species. Laboratories routinely culture viruses specific to humans but they do not use human cells for this purpose. Cells of quite different species are used, sometimes as widely diverse from human cells as the cells of a chicken embryo. The curious point therefore arises that viruses are not specific when they multiply within individual cells. It is their attack on whole animals that appears to be specific. One might wonder what the difference might be. The answer must lie in the operation of the immunity system of the whole animal. The specificity of viruses in most cases is in relation to immunity systems, not to cells. This suggests an inversion of the usual way of looking at viral disease. Instead of thinking of viruses outwitting cells, which is not really possible because of the exceeding paucity of the genetic material in many viruses, we should think of viruses being "invited" into our cells. We should think of our immunity system as constantly scanning the newcomers, with a view to permitting our genome to seize upon any that might be valuable from an evolutionary point of view. Those that are clearly useless are excluded. Those with possible potential are encouraged to interact with the cells, with different animals scanning different viruses according to their separate needs. Only if a virus bears a promise of being useful to the host is it permitted to attack. This view goes against the grain of normal thinking since we inevitably tend to consider viruses as bad because we suffer individually from them. However, the suffering of an individual is irrelevant to evolutionary biology. What matters is the occasional success, not a million failures.

As we mentioned earlier the first life-bearing comet to arrive at the Earth and to find our planet congenial to life was one that brought living cells some 3.83–4 billion years ago. The arrival of primitive life forms could not have stopped in the distant past. Bacteria, larger cells and fragments of cells and viruses and viroids have, on our point of view, continued to arrive through cometary injections right up to the present time. It has been this steady arrival of such genetic material from space that has led to a progressive evolution of terrestrial life.

With each arriving comet, viral particles (probably derived from cells) spanning an astronomically wide range of properties could interact with life forms already present on the Earth. Some viral genes simply add on to host cell lines, others have the effect of attenuating life forms that become, as we say, "diseased".

No amount of shuffling of the genetic material of a primordial bacterium could by itself lead to a flowering plant on the one hand, and a human on the other. As we have already stressed evolution is a fact that cannot be denied, but according to the present point of view, evolution in the way we find it could only have taken place if it was driven from outside — from the wider cosmos. It is in the external universe that all the transformative genes reside, embracing the widest range of possibilities. These would be in the form of mobile genetic units that have been swapped around on a cosmological scale for over 10 billion years.

Sudden surges in evolution and diversification of species of plants and animals, and the equally sudden extinctions evident in the fossil record, point to sporadic additions of genes possibly linked to the arrival of new comets. Cometary genes could in this way become grafted onto pre-existing biological stock leading to the emergence of dramatically new lines. At the same time the effects of epidemic disease could lead sporadically to extinctions over relatively short timescales.

A dramatic example is provided by the extinction of the dinosaurs 65 million years ago. These highly successful reptiles that dominated the planet for over 100 million years disappeared in a very short period of geological time. We know that a comet or comets were somehow involved because of the unusually high content of the element iridium that was discovered in the sediments where dinosaur fossils were found. It seems likely that this major extinction event was caused at least in part by a purely physical process such as the blocking out of sunlight by cometary dust. There is evidence showing a wide range of terrestrial life forms becoming attenuated for a few thousand years on either side of the 65-million-year boundary (so-called K–T boundary). A protracted episode of viral infections from space may have been a contributory factor.

Another important event that took place simultaneously was the great surge of mammals that was to lead in due course to the emergence of Man. Although some evidence of mammalian characteristics can be found at earlier times, the main expansion of mammalian orders occurred 65 million years ago. Both the extinction of some orders and the emergence of others on a massive scale points to a major genetic storm that came with the arrival of cometary viruses 65 million years ago.

Looking across the fossil record in its entirety one sees that this type of event is by no means unique. The fossil record can be seen as largely static with no evolutionary changes occurring except at several sharply defined moments in geological time. As with dinosaur extinction, such changes as occurred are not confined to individual genera or small groups of related genera, as one might expect on the standard picture of Earth-bound evolution, but they extend over a broad sweep of phyla and orders all at once. Evolutionary biologists have described these events as implying an evolution by "punctuated equilibrium", although this name itself offers no explanation of the cause of the so-called "punctuation". On the standard Earth-bound picture such phenomena are not easy to explain.

Hoyle and the present author, in our book *Evolution from Space* first argued that sequential copying of genes would accumulate errors, but on the average such errors must lead to a steady degradation of information, at any rate in simple systems that lack bisexual protection. It is not easy to prove convincingly that the information provided by one single cell with some 500 genes can be upgraded by sequential copying to produce a human with 25,000 genes, and all other living things that inhabit our planet in the absence of a steady input of pristine genes. The processes of mutations, gene doublings and natural selection can only produce at best minor effects as a kind of fine tuning of the whole evolutionary process.

There is, in the view of the author, an absolute need for a continual addition of new information for life, an addition that extended in time throughout the entire period of the geological record.

We mentioned earlier that if comets brought the first life to Earth 4 billion years ago, the additions of pristine microorganisms and new genes (including viruses) from comets must have continued throughout geological time, and consequently played a role in the process of evolution. Such considerations can be extended to the model discussed earlier where genetic products of local evolution on a planet like the Earth were distributed and mixed on a galactic scale. In this context it should be noted that even partially destroyed DNA strands belonging to life forms evolved locally could carry the information of life far and wide (Wickramasinghe, 2010). In this picture impact episodes and gene distribution events would happen recurrently whenever the Oort cloud of comets surrounding our planetary system is disturbed by the gravitational effect of a passing interstellar cloud.

Since we cannot consider the Earth and our own solar system to be unique in this regard, it has to be assumed that similar gene dissemination processes operate for every life-bearing planet in the galaxy. The stochastic nature of gene acquisition events resulting from encounters with molecular clouds leads naturally to a stochastic component of biological evolution — *e.g.* sudden jumps, as we have noted in the Earth's record of life.

It is a consequence of the process we envisage here that many aspects of biology that are introduced in this manner would give the appearance of being pre-programmed. This would happen when the relevant transported genes had evolved over vast timescales and in very many locations and suddenly came to be expressed locally on the Earth. The evolution of the eye may be seen as one example of this type, and even some highly complex, and less definable manifestations of gene expression in our own immediate line of descent in hominid evolution bear the signs of "pre-programming" or pre-evolution. The Japanese Biologist S. Ohno (1970) wrote thus:

"Did the genome of our cave-dwelling predecessor contain a set or sets of genes which enabled modern man to compose music of infinite complexity and write novels with profound meaning? One is compelled to give an affirmative answer...... It looks as though the early *Homo* was already provided with the intellectual potential which was in great excess of what was needed to cope with the environment of his time…"

From the foregoing discussion it is clear that genomes of extraterrestrial viruses and bacteria that are sometimes responsible for pandemics of disease became incorporated in the germ lines of survivors and provided the main driving force for biological evolution (Hoyle and Wickramasinghe, 1979, 1980). Although we were criticised at the time for advocating what appeared to be a return to a primitive superstition with regard to comets and plagues, advances in molecular biology in the new millennium are beginning to provide evidence in our favour. Of particular importance is the discovery of horizontal gene transfer as a process operating across a wide range of phyla, and one that could readily be seen to extend over a galactic or intergalactic scale.

The cosmic theory of life *requires* that genes which are the products of evolution in some distant cosmic location can, on occasion, be transferred to evolving life forms on the Earth (Hoyle and Wickramasinghe, 1982). In this way evolutionary advantage or novelty could be acquired by terrestrial organisms on a stochastic basis, whenever alien genetic material carrying new information is introduced to the Earth and becomes available to terrestrial biology. This in fact constitutes an astronomical process of *horizontal gene transfers* — transfer of genetic information across normal mating barriers taking place on a cosmological scale. It is only in this way that "pre-programmed genes" can make their sudden appearances as we have seen.

The mechanism of Earth-bound horizontal gene transfer (HGT) has now been amply documented (Keeling and Palmer, 2008; Boto, 2010). There is compelling evidence to support a once contentious view that HGT provides an important source of new genes and functions to recipient organisms and is also a driving force for evolution. It has also been recognised that it is the operation of horizontal gene transfer that has foiled attempts to reconstruct ancient phylogenetic relationships in the search for a Last Universal Common Ancestor (LUCA) in the tree of life (Jain *et al*., 2003). It is becoming amply clear that there was probably no such entity localised on the Earth, but rather a cosmic ensemble of genes that has an antiquity comparable perhaps with the age of the Universe itself (Joseph and Wickramasinghe, 2011; Gibson *et al*., 2011).

From all the available data we can infer that sudden shifts in evolution, including the emergence of new species, occur through horizontal gene transfers rather than by the slow neo-Darwinian process of mutations and natural selection. Although the occurrence of Earth-bound neo-Darwinian evolution is not denied, it would probably be dwarfed by interstellar horizontal gene transfers in the long term. The phenomenon of "punctuated equilibrium" that we have already alluded to, where long periods of evolutionary stagnation are punctuated by sharp episodes of innovation and progress, is consistent with cosmically mediated gene transfers. The long periods of slow evolution are due, on the other hand, to Earth-bound neo-Darwinian processes where no external gene inputs are involved.

The successful transfer of genetic information from one organism to another — on the Earth or even across the galaxy — in a manner that permits transmission into a host's germ line requires a vector. The vector could take the form of a plasmid, virus or a bacterium, in which case the host cells and vector are required to enter into some form of symbiosis. Eukaryotes possessing mitochondria or chloroplasts provide living evidence of such horizontal gene transfers that happened in the distant past, with mitochondria and chloroplasts being prokaryotic endosymbionts. Similar symbiotic accommodation of viral genes may have occurred repeatedly in the primate ancestral line that led eventually to *Homo sapiens*.

Sequencing the human genome has been one of the most outstanding scientific accomplishments of the new millennium. It has led to a wide range of discoveries that are transforming our ideas about viruses, disease and evolution (Venter *et al.*, 2001). One surprise was that the total number of genes in human DNA (sequences coding for proteins) was as small as 20,000–25,000, rather than over 100,000 as had hitherto been suspected. Another surprise was that a large fraction of our DNA consists of sequences attributable to viruses, mostly endogenous retroviruses — RNA viruses that reverse transcribe their RNA into DNA. Their significance in causing disease as well as contributing to evolution is only just coming to be understood, and many astounding correspondences with predictions from the theory of *Evolution from Space* (Hoyle and Wickramasinghe, 1980) cannot be overlooked.

The new evidence from genome sequence studies points to frequent episodes of so-called retroviral infections (of which HIV is an example) not only in humans, but in almost all mammalian species. De Groot *et al.* (2002) have identified an entire repertoire of genes known as MHC class 1 genes in chimpanzees, and these genes confer immunity against chimpanzee-derived simian immune deficiency virus. The inference is that modern chimp populations represent descendants from the survivors of a HIV-like pandemic that very nearly culled the entire ancestral chimp line in the distant past. The Hoyle–Wickramasinghe conjecture that HIV was an invader from space was much ridiculed when it was first proposed, but recent developments would appear to restore it at least to the realm of reasonable hypothesis.

Following the integration of a retroviral gene sequence into a host's DNA, random mutations and the development of host immunity leads within a few generations to the cessation of infectivity. This may well be the long-term fate of the human HIV virus in the absence of any artificial medical intervention. Survivors of such major pandemics would thus carry DNA sequences of retroviral origin reflecting the history of prior infections. These viral gene sequences could then conceivably contribute to evolutionary potential in the long term (Hoyle and Wickramasinghe, 1981, 1982).

The process by which viruses are endogenised and included in host genomes is not confined to retroviruses. A non-retrovial RNA transcript appears to have been incorporated in the germ line of several mammalian species, including rodents around 40 million years ago (Horie *et al.*, 2010).

Bacterial infection can also leave an imprint on genes. In a recent article Wang *et al.* (2012) have shown that two immunomodulatory genes called SIGLEC related to bacterial infection are inactive in humans, but not in related primates. The conjecture is that these genes when they were fully active could have been targets for a lethal bacterial infection that nearly culled the human population in the past, perhaps 100,000 years ago.

Summing up: The evidence that now exists for purely Earth-based gene transfers can be extended to transfers over a galactic scale. Transfers of alien genes in the form of viruses would take place whenever the solar system (and the Earth) encounters genetic material (viruses and bacteria) from comets or planetary systems from which such material had been expelled. The appearance of seemingly pre-programmed developments in biological evolution, including the incorporation of such elements well ahead of their utility (*e.g.* intellectual potential in humans), can be readily understood on this basis. Carl Woese' "tree of life" would then be a trivial reconstruction on the Earth of a cosmically derived evolutionary scheme.

Chapter 7

Evidence from Epidemics

Our ancestors of a distant past were apparently unanimous in the belief that comets were the cause of disease and pestilence. The association of comets with epidemic disease appears to have crossed the boundaries of geography and culture. We tend nowadays to dismiss such ancient ideas as primitive superstition born out of ignorance. But was this really so? Our forebears were probably more objective and pragmatic in the way they looked at the world. They were not bound by adherence to dogma or hindered by sociological constraints as we seem to be. Nor were they constrained by the authority of institutions that decided what was respectable and proper to believe and what was not. So any ancient assertion that one natural phenomenon A is linked to another B should perhaps be given credence, at any rate as a working hypothesis worthy of further exploration. The implication would be that a correlation between phenomenon A and phenomenon B had been verified over many generations and this information transmitted through oral tradition. In general it would follow that the older a tradition is the greater is the likelihood that it has a factual basis, and the more respect it therefore deserves.

The connection between comets and epidemic disease cannot be properly understood independently of the wider issues relating life to comets. We argued in earlier chapters that life could not have originated *de novo* on the Earth, and the picture that emerged was that interstellar space is filled with living cells, cells that are mainly in a frozen dormant state.

In our own solar system, we argued that life in the form of bacterial cells was first housed in the comets. Every one of nearly a hundred billion comets forming the Oort cloud at the beginning had warm watery interiors and all the organic nutrients necessary to make for a congenial breeding ground for microbial life. We mentioned in earlier chapters that comets crashing on to the Earth brought the oceans and the atmosphere. With the oceans and the atmosphere so deposited, comets also seeded our planet with life — a life that was able to take root and flourish under the protected canopy of cloud-covered skies.

The first successful seeding of the Earth with life occurred at about four billion years ago. But according to this picture the process of cometary seeding could not have stopped at a distant geological time. The Earth today is well and truly entwined in the debris that is shed from comets. This is illustrated in Fig. 7.1 which shows the projections of the orbits of short-period comets lying inside the orbit of Jupiter.

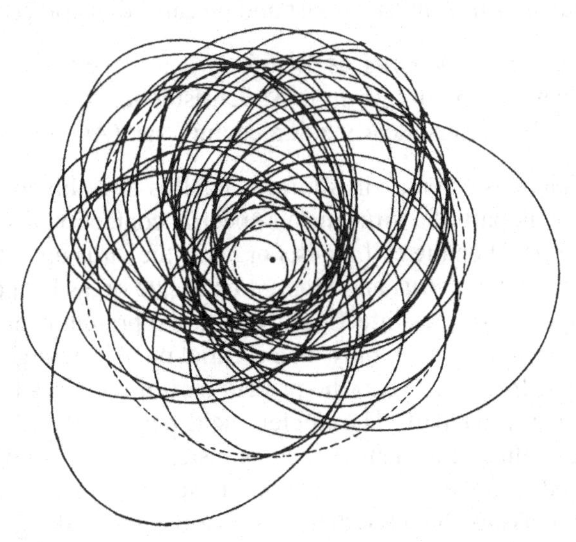

Fig. 7.1. Orbits of short-period comets projected onto the ecliptic plane.

On the average about 100 metric tonnes of cometary debris enters the Earth's atmosphere every day, of which a large fraction is organic and, in the present author's view, biological. Much of this debris is either sterile due to exposure to the Sun, or is burnt up on entering the Earth's atmosphere. But it is inevitable that a small fraction of the incoming dust, freshly evaporated from comets, would contain microbes — bacteria and viruses — that actually survive entry through the Earth's atmosphere. If this fraction is as small as $1/10^{th}$ of a percent, 0.1 tonne per day of viable microbial material will be arriving at the Earth. The total number of bacteria thus arriving per year could be as high as 10^{21}. And the total number of viruses could number upwards of 10^{24} per year, exceeding the number exuded by the entire human population of the Earth by several powers of ten.

Viruses occupy a grey area in biology between living and non-living states. Its essential components as was noted in Chapter 6 involve a protein coat or capsid enveloping a genome comprised of either DNA or RNA that code for function. A virus can replicate only within the cell of a host it infects, and such infections are known to cause a large number of diseases in plants and animals. Examples of human diseases caused by viruses include the common cold, influenza, SARS, smallpox, polio and HIV. In our model of cosmic life, genes of eukaryotic cells as well as their associated viruses must coexist and be carried in comets.

The behaviour of a virus as it infects a host cell cannot be better described than the account given in Sir Christopher Andrewes' book *The Common Cold* published nearly half a century ago (Andrewes, 1965):

"What happens when a virus infects a cell is probably something like this. The protein part of a virus makes specific contact with something on the cell-surface. Then the cell ingests or takes the virus up within itself. It may ingest the whole virus and break it up inside or, as happens with the bacteriophages or viruses infecting bacteria, the protein coat of the virus may be left outside the cell, only the essential nucleic acid gaining access to the interior. In either event, the protein part of the virus is expendable and plays no further part. The nucleic acid part, however, proceeds to instruct the cellular mechanism in a sinister manner. Suppose it is a rhinovirus infecting a cell lining your nose. The instruction will run thus: 'Stop making the ingredients necessary for making more nose-cells. Henceforth use your chemical laboratory facilities for making more nucleic acid like me.' The intruding virus nucleic acid gives the further instruction: 'And now make a lot of protein of such-and-such composition which 1 require wherewith to coat myself'. The cell can do nothing but obey and as more new virus particles are thus assembled by the cell's chemical mechanisms they are, at the end of the production line, turned out into the outside of the cell. With many viruses, including probably rhinoviruses, the final effect is to exhaust the cell altogether, so that after a while it dies and disintegrates. The virus set free will infect more of its victim's cells until such time as defence mechanisms have been mobilised. It will also get into the outside world and infect more victims, for one result of the cell-destruction in the course of a cold infection will be inflammation, pouring out of fluid, sneezing and spread of virus. All of it a very conveniently organised affair for the benefit of the virus…"

As we mentioned earlier, viruses are known to be able to swap genetic information between species (horizontal gene transfer), and this property is used nowadays for medical applications as well as in genetic engineering such as the production of genetically modified crops. Viruses could also add naturally to the genomes of creatures they infect. It is of interest in this context that eukaryotes, including human cells, have substantial stretches of viral DNA that is normally unexpressed. The silent viral DNA plays a crucial part in the evolution of species as was discussed in Chapter 6.

Viruses are associated only with eukaryotic cells of higher life forms. The corresponding infective agents of bacteria are called plasmids, and these could also exchange genetic information between species. The interaction of a plasmid with a bacterium could, for instance, transform a non-pathogenic variety of the bacterium into a pathogenic form. Pathogenic strains of the normally benign *E. coli* (*E. coli* 01570) that appears sporadically and mysteriously from time to time may well be produced in this way, with the culprit transformative plasmid possibly coming from outside the Earth.

Until recently it was thought that the total number of distinct bacterial species on the Earth was a mere few thousand. Today, using modern techniques to detect sequences of nucleic acids rather than intact bacterial cells, it is estimated that the total number of bacterial species might be well in excess of a billion. The vast majority of these bacteria are thought to be what are called extremophiles, bacteria seeking extreme and hitherto unchartered environments. They have not yet been cultured, perhaps never will. They lie in surface soil and surface water, evidently doing nothing, perhaps they are waiting for the right host to emerge. The possibility that they are falling from the skies is supported by the balloon experiments that will be discussed in a later chapter.

A more direct way of testing space incidence arises if this microbial input includes microorganisms that are pathogenic to plants and animals.

Rather as physicists use amplifying electronic counters to detect very small fluxes of incoming cosmic ray particles, so in a similar way plants and animals could be regarded as amplifying detectors for microbes from space.

In general, an incoming pathogen (virus or bacterium) from space has one of two logical paths to follow. It can either establish a reservoir in some group of plants or animals, and thereafter proceed to propagate by horizontal transmission; or it could have the property of not being able to form a stable reservoir. In the first case, of which smallpox and HIV may be examples, epidemics would arise whenever a reservoir effectively breaks its banks. In the second case, every epidemic, every single attack, must be driven directly from space. We shall see later in this chapter that this may well be the case for influenza.

A reading of medical history provides ample evidence for such invasions from space and of a changing pattern of human diseases. Many bacterial and viral diseases have a record of abrupt entrances, exits and re-entrances on to our planet, exactly as though the Earth was being seeded at periodic intervals. In the case of smallpox, a disease caused by a virus, the time interval between successive entrances judged from historical and archaeological evidence seems to have been about 700–800 years. Since man is the only host of the smallpox virus, global remissions of this disease lasting for many hundreds of years are very hard — almost impossible — to understand. From the conventional Earth-centred point of view one has to say that the virus became extinct, and then re-evolved to precisely its original form from some unknown ancestor after many hundreds of years. A rather improbable contingency that must be. From skin lesions discovered in Egyptian mummies one could argue a case for smallpox being prevalent a few hundred years before the classical period in Greece. More decisively, perhaps, we could argue that it was present a few hundred years after the dawn of the Christian era. With the accurate medical records that are available at the time we can be equally sure that smallpox was *not* present in classical Greece and Rome. In fact the Latin word for pock mark, *variola*, appears to have been coined only in the 7th century AD, again supporting the view that smallpox was absent in the early Christian era. A genuine world-wide absence would seem to be implied at this time, for otherwise it would be hard to imagine a disease so infectious as smallpox being kept out for so long from the Western World.

A bacterial epidemic disease that followed a similar pattern of recurrent entrances and exits over the centuries is the bubonic plague. The first thing to note here is that the bubonic plague is not primarily a disease of man. The bacterium *Pasteurella pestis* which causes the plague attacks many species of rodent, as its first target. Because the black rat happened to live in close proximity to humans, nesting in the walls of houses, the physical separation of people from the black rat was always small and could be bridged by fleas. Fleas carried the bacterium from rats to humans (Hoyle and Wickramasinghe, 1979).

Although the transfer of the bacterium from human to flea and back to human presumably occurred, it does not seem to have been sufficient to maintain the disease. Every epidemic of the disease died out as the supply of rats became exhausted. As with smallpox, bubonic plague has come in sudden bursts separated by many centuries, and there are the same difficulties of understanding where *Pasteurella pestis* went into hiding during the long intermissions. A somewhat ambiguous reference occurs in the Old Testament, at a date of about 1200 BC, when the Philistines are said to have been attacked by 'emrods [buboes] in their secret parts...' as a reprisal of God for an attack on the Hebrews. A clear reference to bubonic plague also occurs in an Indian medical treatise *Charaka Samhita* written in the fifth century BC, in which people are advised to leave houses and other buildings "When rats fall from the roofs above, jump about and die."

There may have been an outbreak of plague during the first century AD, with centres of the disease in Syria and North Africa, but between the first and sixth centuries there were no known attacks of the disease. In 540 AD, a pandemic involving the Near East, North Africa and Southern Europe is said to have had a death toll that reached 100 million, with more than 5,000 dying each day in Constantinople alone. This epidemic was the so-called Plague of Justinian, Justinian being the Roman Emperor of the time.

Bubonic plague would then seem to have disappeared from our planet for eight whole centuries, before reappearing with devastating consequences in the Black Death of 1347–50. Thereafter, the disease smouldered with minor outbreaks until the mid-seventeenth century when for two centuries it seemed once again to have died out, only for it to reappear in China in 1894. In India, it killed some thirteen million people in the years up to the First World War.

If this recurrent pattern of pandemics is to be explained on the basis of *Pasteurella pestis* being an endemic microbe, its ability to cause pandemic disease at widely separated epochs could possibly be explained on the basis of a space-incident plasmid or virus that transforms a harmless bacterium into a highly pathogenic form. Without such an effect the available facts are difficult to explain.

There are also many instances of geographically localised attacks of epidemic diseases in historical times. A great plague broke out in the ancient city of Athens in the year 430 BC. The Pelopponesian War that marked the decline of Athens commenced a year before. The chronicle of this war was written by the historian Thucydides with what has been described as minute and scientific accuracy. He describes the plague in such great detail that many a modern physician has tried to identify the disease from clinical symptoms. This has proved exceedingly difficult. Some have tried to link the disease to smallpox, but in this case it should have spread like wild fire and not have ended as abruptly as it did. After reviewing various alternative possibilities one medical commentator of the last century wrote thus:

"I have looked into many professional accounts of this famous plague, and writers, almost without exception, praise Thucydides' accuracy and precision, and yet differ most strongly in the conclusions they draw from the words. Physicians — English, French, German — after examining the symptoms, have decided it was each of the following: typhus, scarlet, putrid, yellow, camp, hospital, jail fever; scarlatina maligna; the Black Death; erysipelas; smallpox; the oriental plague; some wholly extinct form of disease……"

Amid this confusion it would be safe to conclude that it was unlike any disease that was known before or after this time. It was localised around Athens, appears suddenly from nowhere, and disappears equally suddenly and mysteriously.

A medical puzzle in more recent historical times involves a group of Trio Amerindians who for a long time had lived in isolation from the rest of humanity. When the forests in South America were cleared in the early part of the last century a tribe of 500 Trio Amerindians were discovered by anthropologists. It was found that there were several victims of polio amongst them and it transpired that they contracted the disease at times coincident with epidemics in cities hundreds of miles away. There was no conceivable way by which the forest dwelling Surinam Indians could have contracted polio from city dwellers. But the city dwellers and the Indians could both have caught the disease if the causative virus (or a trigger for it) rained on them from above.

A modern bacterial disease that is also difficult to explain except by a component or a trigger that is vertically incident is whooping cough or *Pertussis*. *Pertussis* has for long been known to occur in cycles of about 3.4–3.5 years, which used to be explained on the density of susceptibles theory, namely that after children susceptible to the disease become exhausted by a particular epidemic it was then supposed to take about three and a half years for new births to rebuild the density of susceptibles to the level at which a further epidemic would run. Thus the periodicity on this theory should have been a function of population density, with the shorter periods being found in inner city areas of very high density. However the period 3.4–3.5 years was found to be everywhere the same, in town and country alike, and from one country to another. If the standard theory had been correct, the sudden reduction in the density of susceptibles brought about in the 1950's by the introduction of an effective vaccine should have greatly disturbed the 3.4–3.5 year periodicity, or even destroyed it altogether. Yet the periodicity persisted exactly as before, but with the total number of cases much reduced.

Perhaps the strongest evidence of all for an ongoing incidence from space relates to the case of influenza. At the end of the 19th century, before the viral nature of influenza was established, the distinguished English physician Charles Creighton asserted that the spread of the disease could not be explained on the basis of person-to-person infection. The data from several epidemics had shown that outbreaks occurred almost simultaneously over great distances, but the spread over localised regions took place in a more leisurely manner.

Perhaps the most disastrous influenza epidemic in recent history occurred in 1918–1919 and caused some 30 million deaths. After carefully reviewing all the available information about the spread of influenza during this epidemic, Dr. Louis Weinstein commented thus:

"Although person-to-person spread occurred in local areas, the disease appeared on the same day in widely separated parts of the world on the one hand. but [*sic*] on the other took days to weeks to spread relatively short distances. It was detected in Boston and Bombay on the same day, but took three weeks before it reached New York City, despite the fact that there was considerable travel between the two cities. It was present for the first time at Joliet in the State of Illinois *four weeks* after it was first detected in Chicago. [*sic*] the distance between those areas being only 38 miles."

During the same pandemic in January 1919 Governor Riggs of Alaska reported to a committee of the U.S. Senate that influenza had spread all over an area of the size of Europe and with only a small thinly spread population of about fifty thousand. This was despite conditions for human travel being worse than anybody could remember:

"The territory has to be reached by dog team. You have the short days, the hard, cold weather, and you only make 20 to 30 miles a day. The conditions are such as have never happened before in the history of the territory…"

Here we have a clear case where influenza was not spread by person-to-person contact. The theory that birds harbouring the new virus may have spread the disease by dispersing it in bird droppings into the atmosphere is also disproved because flocks of birds did not fly over Alaska during the freezing winter in November and December. However the winter jet stream could well have overturned the upper atmospheric air, causing a cloud of virus-laden particles to descend on even so large an area as Alaska. Then again the disease could not have been spread from Boston to Bombay in a day at a time when there was no air travel. Nor could the fastest flying birds make that journey in time. Nor are there winds that blow over such a route, with such a speed to cover this distance in a day! But a virus or a trigger for it embedded in micrometre-sized particles, falling through the high atmosphere, will in general arrive at ground level at different places at different times. There will always be a place where it arrived first, and this could be where a new epidemic was seen to start. That a space-borne pathogen could touch down simultaneously at two such widely separated places as Boston and Bombay does not strain credulity at all in such a picture.

Thirty years on, and it was the same story all over again. The world-wide epidemic of 1948 apparently first appeared in Sardinia. A Sardinian doctor, F. Margrassi, commenting on this writes:

"We were able to verify the appearance of influenza in shepherds who were living for a long time alone, in solitary open country far from any inhabited centre; this occurred absolutely contemporaneously with the appearance of influenza in the nearest inhabited centres."

One of the most striking features throughout this whole story is that the technology of human travel has had no effect on the way that influenza spreads. If influenza is indeed spread by contact between people, one would expect the advent of air travel to have heralded great changes in the way that disease spreads across the world. Yet the spread of influenza in 1918, before air travel, was no slower and no different from its spread in more recent times.

It is commonly held that person-to-person transmission of influenza is proved by the very high attack rates in institutions such as army barracks and boarding schools. In the early months of 1978 Fred Hoyle and the present author conducted a survey of boarding schools in England and Wales during an epidemic of the so-called Red Flu — a virus type H1N1 that had been absent in the human population for some 20 years — and was suddenly raging at this time. Children under the age of 18 could surely not have encountered this virus during their life time and were therefore equally susceptible to the disease. We saw these conditions to be ideal for testing the hypothesis that the virus or some biochemical trigger for it was falling through the skies. Indeed between 1978 and the present day similar conditions were not repeated. Our sample involved a total of more than 20,000 pupils with a total number of victims of some 8,800. The distribution of attack rates showed wide variations from the average 44% attack rate. Very many schools showed much lower attack rates and only three schools out of more than a hundred had attack rates in excess of 80% that have been claimed to be the norm.

If the virus responsible for the 8,800 cases were passed from pupil to pupil, much more uniformity of behaviour would have been expected. From the spread of the attack rates across the country we saw evidence for great diversity, with a hint that the attack rate experienced by a particular school (or a house within a school) depended on where it was located in relation to a general infall pattern of the virus or its trigger onto the ground. The details of this infall pattern would of course be determined by local meteorological factors. The infall clearly displayed patchiness over a scale of tens of kilometres, the typical separation between the schools.

One of the schools in our survey, Eton College, had conditions that were particularly well suited for testing the doctrine of person-to-person transmission. There were 1,248 pupils distributed in a number of boarding houses and the total number of cases across the whole school was 441. The actual distribution of cases by house is shown in Fig. 7.2.

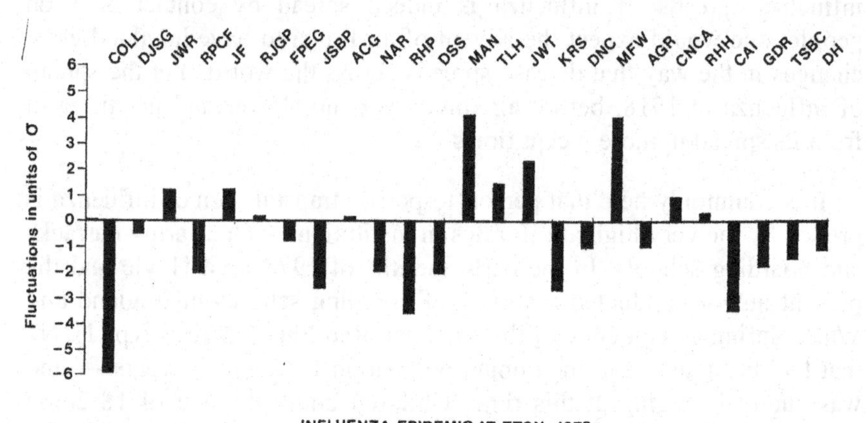

Fig. 7.2. Fluctuations from expected mean numbers of victims in several houses of Eton College expressed in units of standard deviation.

College house with a total population of 70 has only one case, compared with the expected value of 25 on the assumption of random distribution, in a person-to-person infection model. Here again we see patchiness, but now on the scale of hundreds of metres. This entire distribution of attack rates would be expected once in 10^{16} trials on the basis of person-to-person transmission. Clearly, if one looks objectively at the facts alone influenza cannot be seen to be transmissible from person-to-person as our present-day scientific culture would have us believe.

In a public lecture that Fred Hoyle gave in 1981 he described the situation thus:

"It is usually supposed that epidemiological proof of this opinion (of person-to-person spread) is given by the high attack rates on people living institutionally as in hospitals, schools and military barracks. The logic of this supposed proof is as obscure to me as it would be to say that when many spectators in a football match are wetted by a sudden rainstorm, the explanation isn't the rainstorm at all but that the spectators took it into their heads to throw water at each other."

Incidence from space also accounts for the otherwise inexplicable phenomenon that epidemics of flu tend in general to occur with a distinct seasonality in widely separated parts of the world. In normal years the Northern Hemisphere influenza season in temperate latitudes tends to peak in the winter months from December to February, whereas in the Southern Hemisphere, for example in Australia the peak is 6 months displaced in June to August. In the tropics, on the other hand, for example in Sri Lanka, influenza can strike at any time of the year and there is no discernible seasonal effect. This is shown for instance in the statistics of 3 countries shown in Fig. 7.3.

When new strains of the influenza virus arise it is mind-boggling in terms of the normal theory to understand how the disease could be contained within the confines of seasonal winter peaks in the two hemispheres, despite the heavy traffic of airline passengers, for instance between Australia and Europe. If a new strain of the virus emerges first in the East one would expect it to cross into Europe with first plane load carrying an infected passenger or passengers.

According to the point of view developed in this chapter, reservoirs of the causative agent for influenza are periodically re-supplied from a comet or comets at the very top of the Earth's atmosphere. Small particles of viral sizes or smaller tend to remain suspended high up in this region for long periods unless they are pulled down into the lower atmosphere. In high latitude countries, such breakthrough processes, where the upper and lower air becomes mixed, are seasonal and occur during the winter months. Thus a typical influenza season in a European country would occur between December and March as we have seen. Frontal conditions with high wind, snow and rain effectively pull down viral pathogens close to ground level. The complex turbulence patterns of the lower air ultimately control the fine details of the attack at ground

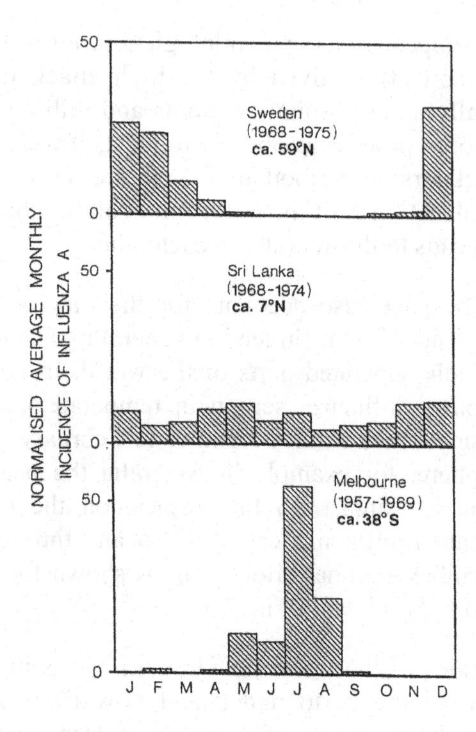

Fig. 7.3. Incidence rates of influenza averaged over several years in three geographical locations.

level, and determine why people at one place and at one time succumb, and why those in other places and at other times do not.

There was a direct demonstration that the general winter down draft in the stratosphere occurs strongly over the latitude range 40° to 60° in the last of the series of atmospheric nuclear tests carried out in the middle of the last century. A radioactive tracer, ^{102}Rh, was introduced into the atmosphere at a height above 100 km and the incidence of the tracer was then monitored year by year through airplane and balloon flights. The radioactive tracer took about a decade to clear itself through repeated downdrafts that occurred with a distinct seasonality. The downdraft was found to be much greater in temperate latitudes than elsewhere, with the period January to March being the dominant months. These observations agree well with the well-known winter season of the viruses responsible for the majority of upper respiratory infections, including influenza.

A possible connection between peaks of sunspot activity and the times of major influenza epidemics, when new viral subtypes were involved, was first suggested by Edgar Hope-Simpson on the basis of data collected over the limited time span 1920–1970. Sunspot numbers give a measure of high-energy activity at the Sun's surface, the peak numbers corresponding in time with frequent solar flares and the emissions of charged particles that reach the Earth. Such activity on the Sun is known to result in geomagnetic storms, ionospheric disturbances that interfere with radio communications, and most spectacularly the production of bright auroral displays, the latter being caused by the streaming of charged particles from the Sun moving along magnetic field lines that connect the Sun and the Earth.

Peaks of solar activity will be expected to assist in the descent of charged molecular aggregates (including viruses) from the stratosphere to ground level. Thus according to our present point of view serious influenza epidemics would follow such peaks, provided the culprit molecular aggregates (viruses) were recently dispersed in the stratosphere from cometary meteor streams. With a more or less regular occurrence of such meteor showers the limiting condition may then be seen as the intensity of solar activity, leading naturally to coincidences between the timings of pandemics or major epidemics and sunspot peaks, as indicated in Fig. 7.4.

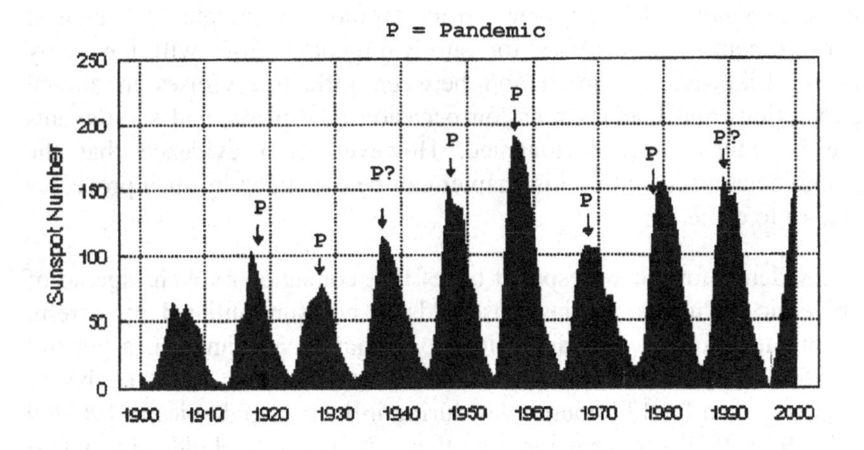

Fig. 7.4. Sunspot numbers in the 20th century compared with positions of major pandemics of influenza.

We have dwelt at length on influenza for the reason that here we might have the most direct evidence of ongoing incidence of biological material from space. It is also the case that influenza poses a continuing threat to humanity, and a future devastating pandemic is seen as a real possibility that needs to be faced. So far all the attempts to deal with this doomsday prospect have met with only partial success. The changing patterns of the dominant virus subtype throughout the 20th and early 21st centuries are depicted in Fig. 7.5.

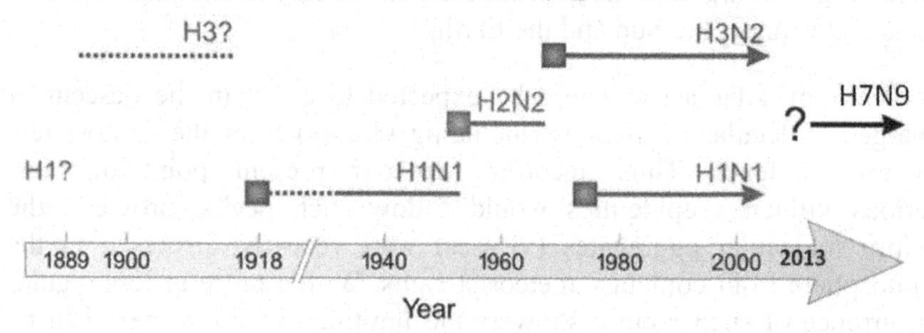

Fig. 7.5. Changes in the dominant influenza virus subtype.

In an influenza virus that is recycled trillions of times through hosts it is inevitable that copying errors would accumulate and against various selection pressures for survival genetic drift will inevitably occur. Likewise, recombination between influenza viruses in animal populations and humans must on occasion take place, and such events are of course well documented. However, firm evidence that the phenomena discussed in this chapter can be explained by such processes is hard to come by.

Evidence for the widespread belief that contagion explains spread of epidemic influenza within households is not substantiated by careful examination of the relevant evidence. Edgar Hope-Simpson, a general practitioner in Cirencester, came to this conclusion after analysing epidemic data for 134 households during influenza epidemics in 1968/69 and 1969/70. His conclusion was that within a household with a first confirmed case of influenza the probability that other family members subsequently succumbed was close to the average attack rate for influenza in the community. In other words belonging to an infected household, despite the intimacy of contact implied, did not lead to a significantly enhanced risk of catching the infection.

We conclude the present chapter by noting that many aspects of bacterial and viral epidemic diseases that have defied understanding in terms of an Earth-centred theory of life may be nearer to solution if we accept that microorganisms are incident from space. The oft-stated criticism that such a proposition is impossible because pathogens and their hosts have co-evolved on Earth has no validity, if evolution itself involved a long succession of endogenisation events involving viruses from space. We noted in the last chapter that a large component of viral DNA in the human genome is identified with endogenised viruses that can provide a store house of evolutionary potential. Recent studies have also shown that evolution from non-human primates to modern humans may have involved a long succession of pandemic infections over millions of years. Each pandemic was a close call to total extinction leaving only a small residual breeding group of immune individuals to carry through the genomic legacy of the evolving line.

In our writings during the 1980's Fred Hoyle and the present author suggested that it would be prudent to maintain a microbiological surveillance of the stratosphere in a search for incoming pathogens so that vaccines may be developed, if necessary, to avert a future devastating pandemic. It might be expected that, in general, weeks to months would elapse between the arrival of viral particles at the top of the stratosphere and their descent to ground level. This would give enough time for action in the event of a potentially lethal pathogen being discovered. The time may well be ripe for instituting such protection protocols before a devastating pandemic provides macabre proof of the theory of cometary panspermia. As Arthur C. Clarke has said the dinosaurs became extinct because they did not have a planetary protection protocol. On the other hand, if they had such a protocol that eliminated the ingress of all new viruses, we humans may not have been able to evolve.

Chapter 8

Microorganisms Entering the Earth

Besides the radiation resistance of bacteria to which we have already alluded, microorganisms are known to withstand an enormous range of conditions, some of which are astonishing if evolution took place in a closed system on the Earth. Microorganisms have been recovered from the most unlikely places on the planet — from the dry valleys of the Antarctic, deep sea thermal vents, from depths of some 8 km in the Earth's crust, thousands of metres below the ocean surface, in tropospheric clouds and in the stratosphere, to name but a few. There is scarcely a niche, natural or man-made on our planet that has not been colonised by some microbial species.

Survival of microorganisms in the abdomen of insects trapped in amber for some 40 million years appears to be well-established (Cano and Borucki, 1995). This timescale of survival has been extended to a quarter of a billion years in the case of a bacterium entrapped in a salt crystal from a New Mexico salt mine (Vreeland *et al.*, 2001). Survival over such astronomically relevant timescales is particularly important for panspermia because the ambient natural radioactivity of the Earth leads to astonishingly high doses of ultra-low intensity ionising radiation (~40 million and 250 million rads) being delivered to these entrapped dormant microorganisms. It can be shown theoretically that exposure to similar doses of low intensity cosmic radiation occurs over hundreds of millions of years under interstellar conditions and thus can lead to substantial rates of survival — exactly analogous to the case of microbes trapped in amber (Wickramasinghe *et al.*, 2010). Direct proof of the survival of bacteria exposed for months to years to radiation in the near-Earth environment has also been demonstrated in NASA's Long Exposure

Facility. Here *Bacillus subtlis* was exposed to the full glare of solar cosmic rays for several months and found to remain viable.

It has been repeatedly stated by critics that viral or bacterial ingress to the Earth in a viable form is impossible, even if such viruses and bacteria actually existed on a cosmic scale. The assertion has been that microorganisms would all be destroyed by heating as they plunge into the Earth's atmosphere. This can be shown to be untrue. Laboratory experiments on the survivability of bacteria with respect to flash heating on atmospheric entry, carried out in the 1980's, showed that heating even to 1,000 degree temperature above absolute zero for a few seconds under dry conditions does not lead to any significant loss of viability (Al-Mufti *et al.*, 1983). It is true that spacecraft re-entering the atmosphere would be heated to the point of sterilisation at its surface, and certain types of cosmic particles *e.g.* meteoroids of sizes of the order of a millimetre, are destroyed by frictional heating. But this phenomenon is sensitively dependent on the angle of entry, size, composition and the degree of fluffiness of the incoming particles.

Fred Hoyle and the present author have argued that clumps of bacteria, individual bacteria, well as viral-sized particles must all survive atmospheric entry to a significant degree (Hoyle and Wickramasinghe, 1979). Survival is also ensured for even the most delicate biological structures embedded within loosely compacted cometary fragments that come to be dispersed within the stratosphere, or even lower down in the atmosphere. In the latter case the deposition of biological material could be highly localised on the surface of the Earth. This could be relevant to the occurrence of highly localised outbreaks of bacterial and viral diseases as we discussed in the previous chapter — *e.g.* the Plague of Athens. It is also of interest to note that fluffy aggregates of siliceous cometary material were recently recovered in meteorite falls in Sri Lanka and they were indeed found to contain fossilised as well as living biological cells. We shall discuss these findings in a later chapter.

Cometary microorganisms reaching the upper atmosphere — say a height of 100 km — begin to fall under gravity but they are quickly sifted according to size. Particles of bacterial size (radii 0.3–1 μm) continue to fall under gravity and could reach ground level in a matter of a year or two. Viral-sized particles become trapped at a height of 20–30 km in a stratospheric trap and further descent is largely controlled by global mixing circuits of the stratospheric air. These circuits have an essentially seasonal character with the potential of bringing down common viruses to ground level in seasonal cycles — as is indeed seen in the patterns of influenza as we discussed in the previous chapter.

The collection of particulate material in the lower atmosphere at heights below 20 km have been carried out from the late 1950's and consistently turned up populations of particles that resemble bacteria and viruses to varying degrees. The Australian physicist E.K. Bigg recovered particles that are similar in external characteristics to microorganisms in the 1960's (Bigg, 1983). More recently D.E. Brownlee has obtained a large collection of particle clumps of cometary origin from 15 km altitude flights of U2 aircraft equipped with "fly paper" collectors for impacting cometary and meteoritic dust (Brownlee *et al.*, 1977). The identification of these so-called Brownlee particles as cometary material was made by using subtle criteria involving isotope analysis. An example of such a particle complex is shown in Fig. 8.1.

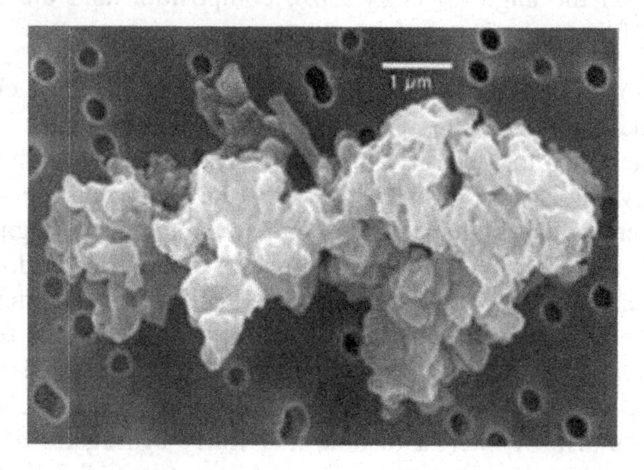

Fig. 8.1. Brownlee particle with carbonaceous chondritic composition.

In all collections of particles from altitudes below 25 km a major difficulty is to distinguish between particles of extraterrestrial origin and those lofted from the surface of the planet. If one flies sterilised balloon-borne packages to sufficiently great heights, however, these difficulties could in considerable measure be overcome. Since vertical movements of the air are in general extremely feeble in the region of the stratosphere, one would not expect particles of bacterial sizes to be carried in air currents to heights much above 15–20 km. Any biological particles discovered at greater heights with equipment that is initially sterilised would therefore have a high probability of coming from space.

Serious attempts to detect microorganisms in the stratosphere were already being made in the 1960's more or less at the dawn of the Space Age. A series of balloon flights into the stratosphere, reaching heights above 40 km, was made by American scientists with equipment to detect bacteria and algae. The results were astonishingly positive in a way that baffled investigators at the time (Gregory and Monteith, 1967). Viable bacteria were recovered that could be cultured by standard techniques. The equipment had been sterilised before each flight, and two identical instrument packages were flown, one of which was exposed to the stratosphere, while the other was not. The unexposed package served as a control. Since bacterial cultures were not recovered from the control package any possibility of laboratory contamination was effectively ruled out.

These early experiments gave results ranging from 0.1 to 0.01 biological cells per cubic metre in the stratosphere, with a density actually increasing with height as the altitude increased in the range of 18 km to 40 km. This is the opposite to the trend one would expect to find for bacteria blown upwards from the ground, and already provided *prima facie* evidence for an ingress of biological particles from space.

In the late 1970's Russian experiments of a very similar kind sought to collect air samples from even higher up in the atmosphere, in the mesosphere, above a height of 50 km. Rockets fired into the high atmosphere expelled detection equipment attached to parachutes. Film was exposed over various height ranges, with particles collected on the film being sealed as the equipment descended out of the height range in question. Recovered film was then examined in the laboratory for microorganisms. After three such flights some 30 cultures were grown of bacteria obtained from heights of 50 km to 75 km. The evidence from both the American and Russian experiments obtained over 4 decades ago would thus seem to favour the idea of bacteria incident from space. If, however, one is impelled through prejudice to regard this proposition as intrinsically implausible, there would be no difficulty in brushing aside the unpalatable results on the grounds of "possible contamination".

The most likely route to ground level for an extraterrestrial microorganism that comes to be dispersed in the stratosphere is via the rain. The microorganisms would effectively serve as nuclei around which particles of water-ice could grow. For many years scientists have been baffled by the problem of how clouds saturated with water vapour come to be seeded so as to produce rain. An atmospheric cloud of saturated water vapour at 0 degree centigrade or slightly lower does not spontaneously turn into rain without either the formation within or the introduction from outside of "freezing nuclei".

Over half a century ago the Australian physicist E.G. Bowen discovered a remarkable connection between such "freezing nuclei" in rain clouds and extraterrestrial particles. He showed there was an astounding link between the frequency of freezing nuclei detected within clouds and the occurrence of meteor showers. Meteor showers occur at regular times of the year as the Earth crosses the trails of debris evaporated from short-period comets. Although larger particles that enter in this fashion would be evaporated quite high in the atmosphere, microorganisms could survive and so be able to act as freezing nuclei. Exceptionally heavy rain was systematically recorded about 30 days after the peaks of meteor activity.

Bowen (1956) wrote as follows in the journal *Nature*:

"The hypothesis has therefore to be advanced that dust from meteor streams falls into the cloud systems of the lower atmosphere, nucleates them and causes exceptionally heavy falls of rain thirty days after the dust first entered the atmosphere."

Although this hypothesis may have appeared far out in 1956, it has since been established that bacteria are often involved in freezing nuclei and that they serve as the most efficient nucleating agents for rain. The 30-day time lapse between the disruption of a meteoroid in the upper atmosphere and rain fall is easily understood as the time of descent of sub-micron particles through the atmosphere.

It was in view of these early results from the 1960's and 1970's that Fred Hoyle and the author tried in the 1980's to persuade the Indian Space Research Organisation (ISRO), through the good offices of our friend Jayant Narlikar, to conduct similar work using their well-attested balloon flight capabilities to sample stratospheric air in search of microorganisms. This project was not considered worthwhile or viable at the time. But after the lapse of several years it was approved and carried out as a joint venture between Indian researchers and workers at Cardiff and Sheffield Universities in the UK (Harris *et al.*, 2002).

Air samples were collected from a balloon flight launched over Hyderabad, India on 21 January 2001 in four height ranges: 19–20 km, 24–28 km, 29–39 km and 39–41 km. The collection involved the use of balloon-borne cryosamplers — a manifold of 16 stainless steel tubes, fully sterilised and evacuated to high vacuum levels. These steel tubes were placed in a liquid neon chamber to cool them to 10 degrees above absolute zero.

The entrance to each stainless steel probe was fitted with a metallic valve which was motor driven to open and shut on ground tele-command. Throughout the flight the probes remained immersed in liquid Ne so as to create a cryopump effect, allowing ambient air to be admitted when the valves were open. Air including aerosols dispersed within it was collected into a sequence of probes during ascent, the highest altitude reached being 41 km. The cryosampler manifold, once the probes were filled with stratospheric air and aerosol particles, was parachuted back to ground.

The air from the exit valve of each probe was passed in a sterile system in a microflow cabinet sequentially through a 0.45 μm and a 0.22 μm micropore cellulose nitrate filters to trap the aerosol particles including biological cells (Harris *et al.*, 2002; Wainwright *et al.*, 2003). Clumps of cocci-shaped sub-micron-sized particles of overall average radius 3.0 μm were discovered from isolates of filters that trapped air collected at 41 km. The clumps were identified first using a scanning electron microscope (Fig.8.2) and subsequently by a technique known as epiflourescence microscopy. The latter involved the use of a membrane-potential-sensitive dye (a cationic carbocyanine) with fluorescence interpreted as revealing the presence of viable cells. Such fluorescent spots are seen, for example in Fig. 8.3.

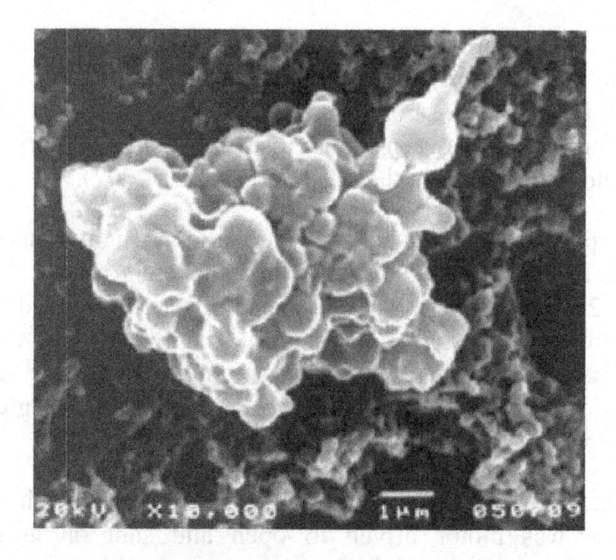

Fig. 8.2. Scanning electron microscope image of clump of cocci and a bacillus.

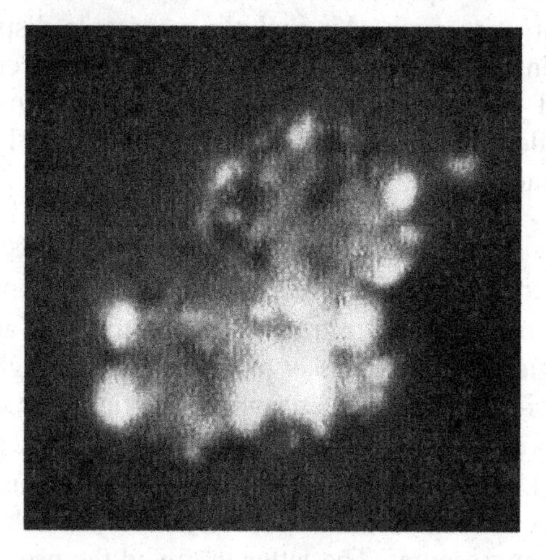

Fig. 8.3. Clump of viable but non-culturable bacteria from 41 km fluorescing on application of a carbocyanine dye.

A similar procedure using the nucleic acid stain acridine orange was also found to reveal the presence of clumps of cells containing nucleic acid. Initial attempts at obtaining cultures were unsuccessful and the cells shown in Figs. 8.2 and 8.3 were deemed to be viable, but non-culturable bacteria. Milton Wainwright subsequently obtained cultures of a coccus and a bacillus from isolates of air filters using a soft potato dextrose agar nutrient medium (PDA) and taking every conceivable precaution against contamination. The cultured bacteria were found to be UV resistant but otherwise similar to well-known terrestrial species.

To most microbiologists the fact that these isolates have similar characteristics to terrestrial microbial species poses a problem, since they tend to *assume* that non-terrestrial microbes if they exist must evolve differently elsewhere. Wainwright's finding is, however, fully consistent with panspermia models in which Earth organisms and evolution itself involves the incidence of cometary organisms on our planet for the past 4 billion years. The main features of all known bacterial genotypes are derived, according to our theory, through a process of cosmic evolution and they are being constantly replenished from space. So to find newly arriving microorganisms genetically similar to their counterpart resident microbes on Earth is to be expected.

With instrumental and laboratory contamination excluded at all stages of the experiment two options remain. Firstly, one might think that the organisms obtained from the stratosphere were carried from the ground in a volcanic eruption or in some other exceptional or rare meteorological event. The other possibility is that they arrived from space. A volcanic origin is ruled out for the simple reason that there was no volcanic eruption recorded in a two-year run-up to the balloon launch date on January 20, 2001, and calculations show that a steady infall would drain out particles of 3 μm radius in a matter of weeks. A similar objection applies to rare meteorological events that may loft such particles from the ground.

Statistical sampling analysis of cell populations collected from a height of 41 km in the stratosphere implied that microorganisms of a presumed cometary origin were incident over the whole Earth at an average rate of 0.1 tonne per day (Wainwright *et al.*, 2003). Critics of panspermia may argue that 3 μm radius particles get burnt through frictional heating and end up as meteors. Some fraction may do so, but most would not. Survival depends on many factors such as angle of entry and mode of deposition in the very high stratosphere. Several modes of entry can be considered that permit intact injection into the stratosphere, possibly starting off as larger aggregates released from comets that disintegrate into a cascade of slow-moving smaller clumps. Evidence for such disintegrations have been available for many years (Bigg, 1983), and more recent studies of Brownlee particles collected using U2 aircraft have also shown the survivability of extremely fragile organic structures that reach the lower stratosphere. Similar particles have also been recovered from Antarctic ice as shown in the example of a carbon rich assemblage in Fig. 8.4.

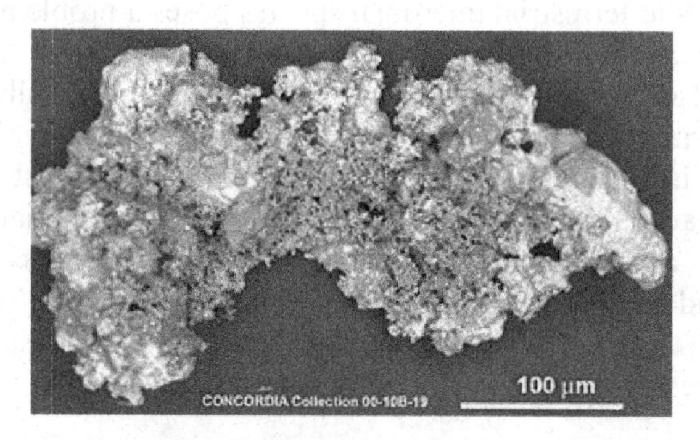

Fig. 8.4. Fragile micrometeorite recovered in Antarctic.

A few years after the cryoprobe experiment of 2001, a second stratospheric aerosol collection from 41 km recovered three new bacterial species with exceptional ultraviolet resistance properties and one of these was named in honour of Fred Hoyle — *Janibacter hoylei* (Shivaji *et al.*, 2009). Of the daily average input into the Earth of some 100 tonnes of cometary material, we can conclude that 0.1 percent is in the form of viable bacteria that reach the stratosphere, and ultimately fall to the surface of the Earth.

Perhaps the most important work relating to microorganisms and biological entities falling from space through the stratosphere was carried out by a team of scientists led by Milton Wainwright (Wainwright *et al.*, 2013a, b, c). A stratospheric balloon was launched on 31 July 2013 from an open field near the town of Wakefield in West Yorkshire. A component of the payload that was ingeniously designed to collect infalling micrometeoroids was a modified CD drawer onto which electron microscope stubs were attached. The sampling drawer with receptive electron microscope stubs was opened for 17 minutes and exposed to the stratosphere as the balloon ascended from an altitude of 22 km to 27 km. The sampling payload was securely sealed and later recovered by parachute, undamaged and intact. A separate control flight with an identical payload was launched into the stratosphere before the actual sampling flight, when the drawer in this case was not opened, but all other procedures remaining the same.

The electron microscope stubs from the recovered payload were removed and examined in a scanning electron microscope. A variety of biological entities of dimensions greater than 30–50 μm were found on the stubs recovered from sampling flight. The fact that no similar structures were found on any of the stubs from the control flight demonstrated that the scrupulous procedures used to prevent ground-level contamination had proved effective.

Two examples of such putative biological structures found by Wainwright *et al.* (2013a, b) are shown in Figs. 8.5 and 8.6. The relatively large 30–50 μm sizes of the structures, and the micro-craters on the stubs associated with some of them, indicate that the particles were falling from space at speed, rather than drifting upward from the ground. The structures in Figs. 8.5 and 8.6 are both distinctly biological; the diatom in Fig. 8.6 is identifiable with a known genus of diatoms.

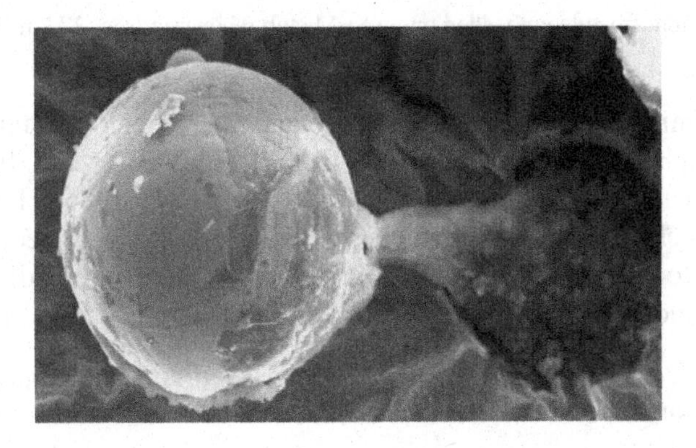

Fig. 8.5. A titanium sphere isolated from the stratosphere with a fungus-like "knitted" cover; the sphere has been moved by micromanipulation across the sampling stub to reveal biological material streaming out and a deep impact crater to the right (Wainwright *et al.*, 2013).

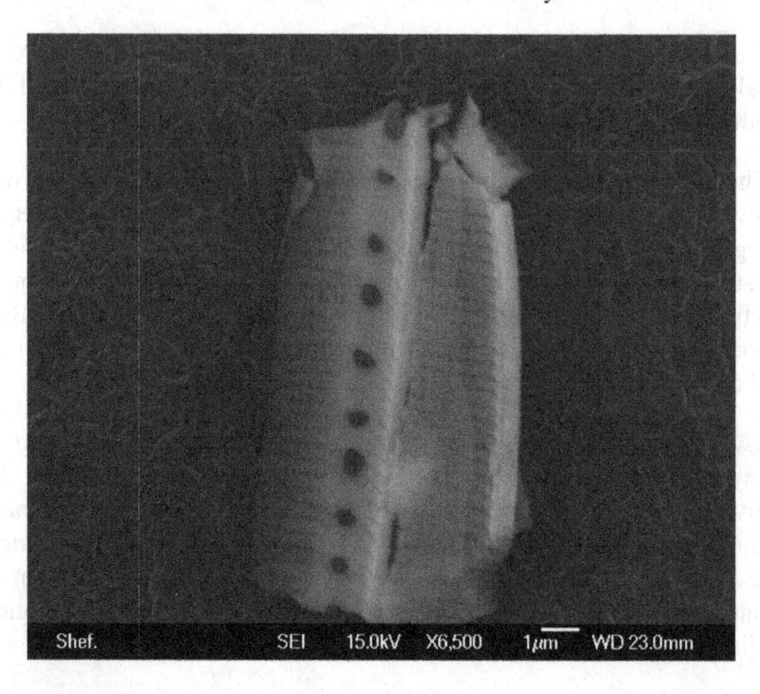

Fig. 8.6. Diatom frustule embedded on an SEM stub recovered from 27 km (Wainwright *et al.*, 2013).

The terminal speed at which a cometary micrometeoroid falls though the stratosphere can be calculated from formulae set out by Kasten (1968) and tabulated atmospheric data (Cole *et al.*, 1965). This falling speed (which scales with the average density of the meteoroid) thus computed for heights of 41, 27, 23 km in the stratosphere is displayed in Table 8.1 and Fig. 8.7 for various particle radii.

Table 8.1. Velocity of falling spherical particles (cm/s; 0.36 km/hr) of various radii, and with density 1 g cm^{-3}.

a/μm / *h*/km	1.0	3.0	10.0	20.0	30.0	50.0	100	200
23	0.048	0.41	4.47	17.8	40.1	111.1	444	1,776
27	0.08	0.68	7.46	29.7	66.8	185.2	740	2,960
41	0.5	4.27	46.6	185.7	417.2	1158.0	4,627	18,500

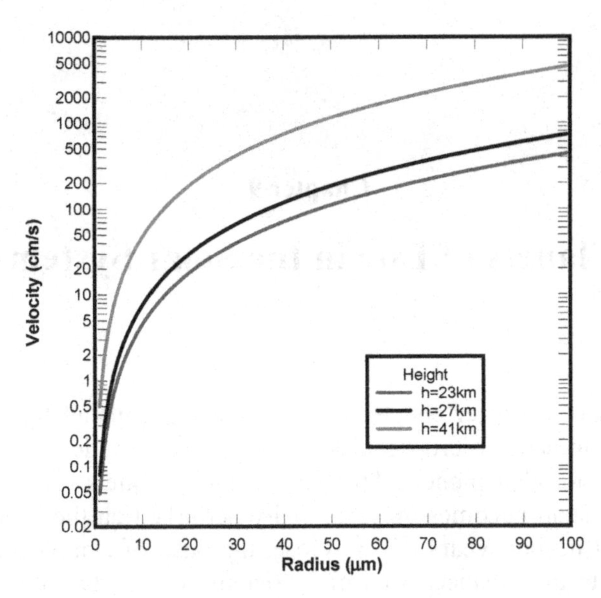

Fig. 8.7. Falling velocities of micrometeoroids of various radii falling through different heights in the stratosphere (average meteoroid density assumed to be unity).

From this calculation we find that a 3 μm radius particle of unit density falls to ground from 27 km in 46.5 days, whilst a particle of radius 50 μm will fall in 4.1 hours. It is thus amply clear that at least the largest of the collected particles in the experiments of Wainwright *et al.* (*e.g.* Figs. 8.5, 8.6) could not be interpreted as a residue from any past volcanic eruption, or the lofting of such particles from ground level over plausible timescales. We also note that the speed at which a 30–50 μm particle falls at 27 km altitude (~50 cm/s) is sufficiently high for the craters such as seen in Fig. 8.5 to be formed.

Chapter 9

Planets of Life in the Solar System

We discussed in earlier chapters that an inhabited planet like the Earth could disseminate microorganisms and even genetic fragments of evolved life to other planets. The Oort cloud of comets surrounding our planetary system becomes gravitationally perturbed on the average once in about 40 million years due to close approach of a molecular cloud. This leads to an enhanced rate of collisions of comets onto the Earth, each of which inevitably leading to the splash back of viable microorganisms into space. Such microbes would be able to infect and colonise newly-forming planetary systems. Within the confines of our own solar system similar impact events can easily redistribute life, as for instance transferring microbial life between Earth and Mars.

Planetary bodies in the solar system have now been explored using a wide range of techniques — optical, ultraviolet and infrared measurements, radar and radio astronomy. But the most dramatic discoveries have come from direct space exploration carried out in the past three decades. In this chapter our main concern will be to review the recent exploration of the solar system and in particular to examine the question whether there is evidence of life, or the potential for life on other planets or satellites in our solar system and beyond.

We have already mentioned that microorganisms have an almost infinite range of survival properties. The recent studies of extremophiles have shown clearly that the limits of life are a rapidly receding horizon. This gives confidence for astrobiologists to search for extraterrestrial life even in locations that might at first sight seem unfriendly or hostile. Within our own solar system a multiplicity of such niches conducive to life surely exists, and the challenge will be to locate them and to discover if life really exists within them.

Mercury

If, in our search for life, we move outwards from the Sun along the sequence of the planets, the first one we reach is Mercury. With a surface gravity about one third that of the Earth and with an orbital period around the Sun of some eighty-eight Earth days, Mercury appears to an observer on the Earth to have phases similar to those of the Moon. From the surface of Mercury the Sun would look nearly three times larger than it does from the Earth, and the solar radiation received on a given area would be nearly 10 times greater than that received by a similar area on the Earth. The period of rotation of Mercury about its axis is about fifty-nine Earth days and the surface temperature varies between about 400 degrees C on the sunward side to about –200 degrees C on the dark side. This intense daytime heat combined with the planet's low surface gravity is the reason why Mercury has an exceedingly thin atmosphere comprised mostly of the heavier rare gases. With a surface which probably resembles that of the Moon, Mercury is perhaps the least life-friendly of the inner planets.

Until recently Mercury would have been totally written off as a home for any form of life. However, new results from NASA's Messenger probe that entered Mercury's orbit in March 2011 has revealed the presence of water-ice and frozen organic materials inside permanently shadowed craters at the planet's north pole — including material resembling tar or coal. Such material was probably delivered by impacting comets and meteorites, and provides tantalising evidence of localised habitats that may be congenial to microbes.

Venus

Moving further out from the Sun, the next planet we come to is Venus. This planet is very similar to Earth in both size and mass, and when they are at their closest the two planets are only some twenty five million miles apart. Venus has an unusual brightness which is caused by the high reflectivity of the thick clouds that shroud its surface, and it has phases like the Moon. At maximum brightness it has a brilliance which is exceeded only by the Sun and Moon. The most accurate information concerning the atmosphere of Venus has come from a series of space probes.

Venus has a thick convective atmosphere dominated by CO_2 producing a powerful greenhouse effect. The average variation of temperature with height is depicted in Fig. 9.1, showing a temperature difference between the surface and the cloud tops of ~500 degrees. The atmospheric composition of Venus shows CO_2 making up 96.5%, the rest including N_2, H_2O, CO, OH, HCl, H_2S, COS and SO_2. Venus's atmosphere has a high opacity to visible and ultraviolet light, reflecting ~80% of the incident solar radiation. Despite the many spacecraft that have visited Venus since the 1970's it is remarkable that the cloud domain of the planet still harbours so many unsolved mysteries. In particular, a complete characterisation of the cloud aerosols that impart a yellow tinge to the clouds still remains uncertain.

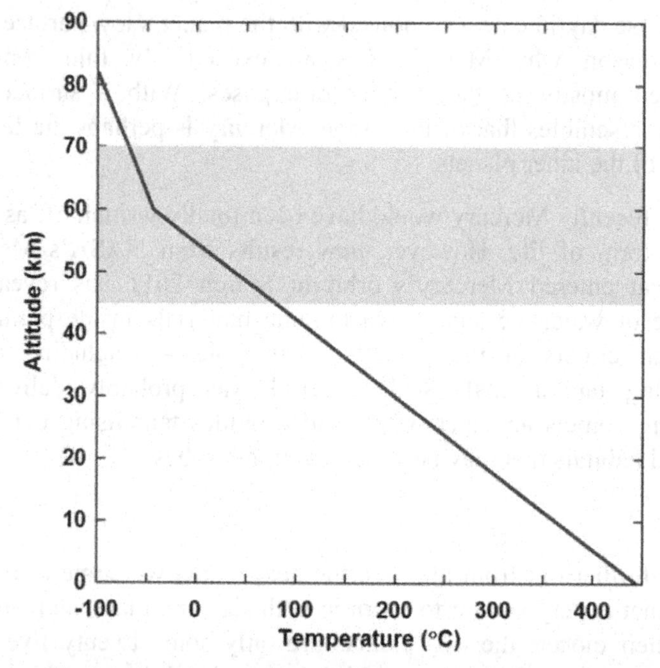

Fig. 9.1. Average temperature in the Venusian atmosphere.

Whilst the conditions near the surface of Venus, T > 460 °C, rule out microbial life, the temperature and pressure regime in the altitude range 45–70 km defines a habitable zone for some types of extremophile bacteria that have actually been found on the Earth. Here the ambient temperature varies between –25 degrees C and +75 degrees C, and the pressure in the range ~0.1 to 10 bars. Speculations relating to a Venus-adapted microbiota have been published over many years (Morowitz and Sagan, 1967; Cockell, 1999; Schulze-Makuch *et al.*, 2004). Water, albeit in small quantities, has been identified in the atmosphere, and this is adequate for microorganisms to concentrate and exploit. Furthermore with a stable cloud system primarily circulating between 70 and 45 km and with a steady supply of nutrients from sublimating meteorites, a Venusian aerobiology remains a distinct possibility (Hoyle and Wickramasinghe, 1982; Wickramasinghe and Wickramasinghe, 2008).

ESA's Venus Express probe (2007) has given us evidence of frequent electrical discharges (lightning) in the atmosphere. Such energetic events would be expected to generate large amounts of CO from CO_2. The absence of higher concentrations of CO than is observed, despite the lightning, might be taken as strongly suggestive of an exotic microbiology. There is a diverse group of terrestrial bacteria and archaea known as hydrogenogens that can grow anaerobically using CO as the sole carbon source and H_2O as an electron acceptor, producing CO_2 and H_2 as waste products (Wu *et al.*, 2005). The presence of H_2S and SO_2 in the atmosphere could also point to the presence of extremophilic "sulphur" bacteria (Cockell, 1999). Droplets of atmospheric sulphuric acid could provide a medium in which acidophiles can thrive. The detection of COS (carbonyl sulphide) in the Venusian atmosphere may also be taken as an indicator of biology. Finally, the sizes and refractive indices of aerosol particles found in the upper clouds of Venus (55–65 km), (Fig. 9.1), are consistent with those appropriate for bacterial spores (Hoyle and Wickramasinghe, 1982).

We envisage a situation in Venus that is analogous to what happens in terrestrial clouds. Sattler *et al.* (2012) have demonstrated bacterial growth in super-cooled cloud droplets, arguing that bacteria in tropospheric clouds are actually growing and reproducing. A stable aerobiology require processes by which (a) bacteria nucleate droplets containing water and nutrients, (b) colonies grow within the droplets, (c) droplets fall into regions of higher temperature where they evaporate releasing spores to convect upwards to yield further nucleation. In the case of Venus this cyclical process would be expected to take place between the tops and bases of clouds.

Mars

On the journey outwards from the Sun, we bypass the Earth and come to the last of the four terrestrial planets, Mars. For a long time it attracted more attention than any other planet because it was considered a likely habitat for intelligent life. With a radius of about half that of the Earth, and a mass of approximately one-ninth, Mars has a surface gravity which is a little less than half that of the Earth. The Martian day is almost exactly as long as an Earth day, and because the tilt of its axis of rotation is the same as that of the Earth, the seasons are also similar to terrestrial seasons. On the other hand, Mars is further than the Earth from the Sun, so that the Martian year is nearly twice as long as the terrestrial year.

Mariner 9 provided the first detailed survey of the whole of the Martian surface photographically with a resolution of one kilometre, and a few percent of the surface at a finer resolution of a hundred metres. Martian craters bear a general similarity to lunar highland craters, except that they are much shallower, and they are almost certainly caused by meteorite impacts. The Mariner probes have showed evidence for volcanic activity at certain sites on Mars and they have given us information about the behaviour of dust storms that occur sporadically over large areas of the surface. The storms give rise to conspicuous changes in the planet's colour when viewed from Earth. A storm which occurred in 1971 was particularly violent and widespread. In certain localised areas such as the "dust bowl" Hellas there may be perpetual dust storms.

The equatorial surface temperature of Mars varies between a daytime high which is close to the melting point of water-ice, and a night-time low of about –100 degrees centigrade. At the two landing sites of Viking 1 and Viking 2 (space probes landing in 1976) the temperatures ranged from a high of –31 degrees centigrade to a low of –84 degrees centigrade. The Viking 2 orbiter recorded a north pole temperature of –68 degrees centigrade. At this temperature carbon dioxide would not be frozen, so that the frozen polar cap material which was seen can be inferred to be water-ice.

The Martian atmosphere is mainly comprised of carbon dioxide (about 95 percent), argon (about 3 percent), nitrogen (1.5 percent) and oxygen (0.15 percent) together with a small quantity of water and traces of methane. The atmospheric pressure at the surface of the planet is about half a percent of that on Earth. Such a thin atmosphere provides no protection from the Sun's ultraviolet radiation, and the flux reaching the Martian surface would prove lethal to most, if not all, terrestrial organisms. On Mars, therefore, life could only exist in subsurface niches where a natural shelter is provided from ultraviolet light. This may not be difficult. Dust basins are possible sites, since their bases may be quite well shielded by the efficient extinction properties of dust. It could be that organic molecules associated with pockets of microbial activity are swept up in dust storms and that the overlying dust protects microorganisms near the surface.

A major goal of the space probes Viking 1 and Viking 2, which landed on Mars on 20 July and 3 September 1976, was to search for microbial life. In one experiment led by Gilbert Levin, biological tests were done *in situ* on samples of soil, some of which were taken from under surface rocks (Levin and Straat, 1976). The presumption was that any microorganisms which may be present had metabolic processes broadly similar to those of terrestrial microorganisms. The soil was treated with various nutrients, and expelled gases and the soil itself were examined in several ways. The results of the experiments turned out to be confusing at the outset, but later convincingly indicative of life. The soil was much more active than any known terrestrial soil and was described by chemists as a super-oxidant. One curious fact is that the bioactivity of the Martian soil apparently persists even after the solid is heated to well above normal sterilisation temperatures. Another remarkable fact is that the Martian soil did not show detectable amounts of organic compounds. This would mean that any Martian microbiology that existed must be remarkably thermophilic (heat-loving) and Martian ecology would have to provide a highly efficient scavenging system for free organic molecules. To err on the side of caution NASA scientists announced in 1976 that the Viking results were inconsistent with life, and that some other explanation was required to explain the vigorous gas release that was observed. As it transpired this was a mistaken judgement that led to a succession of wrong decisions being made in the later planning of Mars missions.

In 1986 and again in 2012 a careful re-examination of all the 1976 data led to the startling (but little publicised) conclusion that the Viking results may have indicated a strong positive result for primitive life in subsurface niches on Mars (Bianciardi *et al.*, 2012). What is true beyond any doubt is that the results of the 1976 Viking experiments remain fully consistent with the presence of microbial life on Mars.

Since 1976 there have been many space missions to Mars. They have found evidence of subsurface water, dried-up river beds as well as methane in the upper atmosphere, all of which suggest that microorganisms could still live in specialised niches close to the surface. And in the distant past, when rivers flowed on Mars, much more abundant life was possible.

In 2004 ESA's Mars Express spacecraft in orbit around Mars detected methane in the Martian atmosphere. The amount of methane found in the high atmosphere cannot be explained in terms of volcanic activity on Mars. With the relatively short dissociation time of this molecule in the atmosphere a steady supply of methane from the surface or subsurface appears to be required, and this also points to extant microbial life.

It is a sad commentary on the sociology of modern science that in the many robotic missions that have landed on Mars since 1976 not a single life-detection experiment was included. There is a touch of irony in that future sample-return missions to Mars, in which rock samples will be brought back to Earth, include elaborate "planetary protection" measures to take care of the contingency that microorganisms might be brought back from Mars — even, perhaps, microbes that may be pathogenic to humans!

NASA's Curiosity Rover equipped with the most sophisticated mobile laboratory landed on the Gale crater of Mars on 6 August 2012 and is billed to spend several years probing for signs of past and even present life. If any indirect evidence for extant Martian life is found it can only be regarded as a long overdue confirmation of a discovery already made in 1976.

Mars Meteorite ALH84001

A further chapter in the exploration of Mars was opened in August 1996 with studies of a 1.9 kg meteorite (ALH84001) which is believed to have originated from Mars. This is one of a group of meteorites discovered in 1984 in Allan Hills, Antarctica, which is thought to have been blasted off the Martian surface due to an asteroid or comet impact some 15 million years ago. The ejecta orbited the Sun until it plunged into the Antarctic and remained there until it was discovered in 1984. The presumed Martian origin of these meteorites (known as SNC meteorites) seems to have been confirmed by several independent criteria. One that is perhaps amongst the most cogent involves extraction of gases trapped within the solid matrix which were found to resemble, in relative abundances, the gases that were discovered in the Martian atmosphere. Also the ratio of oxygen isotopes $^{17}O/^{18}O$ in the mineral component matches the value appropriate to Mars so closely that there is no reason to doubt a Martian origin.

A team of NASA investigators led by David S. McKay (McKay *et al.*, 1996) have found that within the meteorite ALH84001 there are sub-micron-sized carbonate globules around which complex organics deposited. Hazy rims around the carbonate globules are thought to be some sort of an organic biofilm that modern colonies of terrestrial bacteria frequently produce. The molecules that have been discovered include polyaromatic hydrocarbons which we saw in an earlier chapter exist in interstellar space and can be identified as characteristic degradation products of bacteria. Moreover, strings of ovoid shaped structures such as one shown in Fig. 9.2 have been considered to be suggestive of fossilised microorganisms — nanobacteria. Amongst these ovoid structures which are about 20–100 nm in diameter, are also minute single crystals of magnetite, similar to structures laid down by terrestrial iron-oxidising bacteria. Another indication that may point to biology is

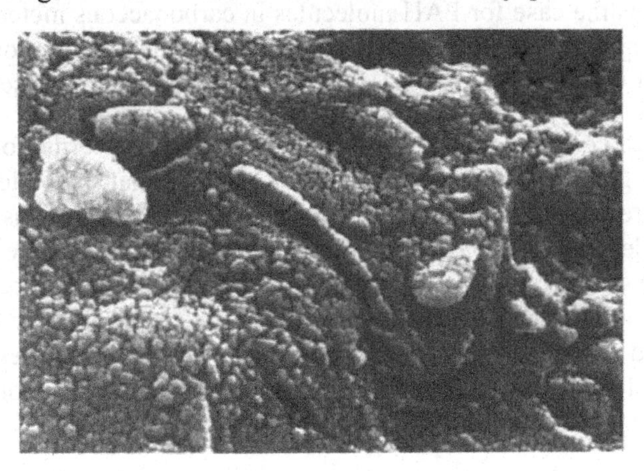

Fig. 9.2. Chain of putative nanobacteria in ALH84001.

an enrichment of the carbon isotope ratio $^{12}C/^{13}C$ that has been found in the carbonaceous component of the meteorite.

An initial worry was expressed in some circles concerning the small sizes of the presumed bacterial fossils in ALH84001. They have dimensions that are 5–10 times smaller than would be expected for normal terrestrial bacteria. However, it soon became clear from the work of Robert L. Folk and E. Kajander that bacteria of similar sizes do indeed exist on the Earth and comprise a large class of organisms known as nanobacteria (Kanjander and Ciftcioglu, 1998). There is considerable evidence to support the view that bacteria with diameters in the range 0.05–0.2 micrometres are largely responsible for mineral precipitation on the Earth.

It would indeed have been surprising if a claim of such profound importance as this is not immediately subject to meticulous scrutiny by the scientific community at large. Less than a year after the original discovery was published, and hailed by President Clinton as the greatest discovery in science, there have been sceptics as well as supporters that have come to the fore. Those sceptics who have voiced the opinion that PAH type molecules could be non-biological have little to back their claim except the fact that interstellar clouds as well as normal carbonaceous meteorites are replete with similar molecules. As we have already mentioned in the context of interstellar dust, these PAH molecules in interstellar clouds also most likely have a biological origin. The same is the case for PAH molecules in carbonaceous meteorites. An abiotic origin of all the organic molecules in meteorites, comets and interstellar space, is a claim that does not stand up to rigorous scrutiny.

Another concern related to the temperature of formation of the carbonate globules in ALH84001, with a claim that estimates of this temperature exceed a survival limit for bacteria. This claim has however been challenged and it can be maintained that the carbonate globules formed at temperatures well within the range normally accepted as relevant for bacterial survival and replication. Moreover, it has been established that ALH84001 had never been heated to temperatures in excess of 110 degrees centigrade since 4 billion years ago, long before any bacteria could have infiltrated the rock.

A further point of contention was the origin of the magnetite particles within the meteorite. A claim that some of these particles are in the shape of whiskers and that they possess crystal defects has been used to argue against their biogenic origin in favour of an origin resulting from condensation in a high temperature vapour. This contention has also been refuted by McKay and his team, and in any case one cannot yet rule out the possibility that biogenically produced magnetite is entirely free of crystal defects.

In July 2011 another meteorite from Mars fell on the desserts of Morocco and was recovered soon afterwards in October 2011 near the village of Tissint. This so-called Tissint meteorite was blasted off the surface of Mars by a comet or asteroid impact several million years ago. A piece of this meteorite was recently examined by Jamie Wallis *et al.* (2012) and found to contain evidence of extinct microbial life. Spherical globules rich in carbon and oxygen were discovered in the interior of Tissint. Figure 9.3 shows one such apparently hollow organic structure that cracked when subjected to a high-energy electron beam in an SEM machine.

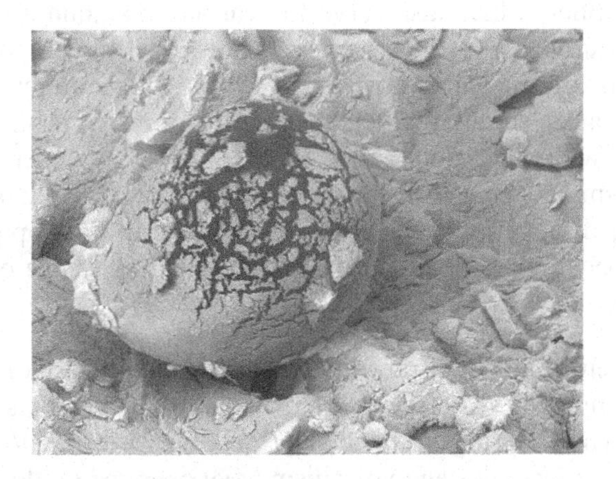

Fig. 9.3. Spherical carbonaceous shell of micron size within Tissint (Wallis *et al.*, 2012).

Further compelling evidence of fossilised microbial structures in the Tissint meteorite has recently been compiled by Jamie Wallis and adds considerably to the contention that Mars was a planet replete with microbial life in its early geological history (Wallis, 2014).

Giant Planets and Moons

When we consider the extremely wide range of conditions that support life on Earth, the prospect for life on the satellites of Jupiter and Saturn cannot be ruled out. The relevant conditions include those prevailing at great depths in the Earth's crust, in the Antarctic dry valleys and in high temperature vents in the ocean floor — conditions that were already discussed in an earlier chapter. Similar conditions must occur quite plentifully in the satellites of the giant planets.

The Jovian satellites were first examined at close range by the Voyager spacecraft in 1979, and they have come under even closer scrutiny following the launch of the Galileo orbiter in December 1997. Io, with a diameter of 5,640 km, has been found to be a hotbed of volcanic activity, some of which could be triggered by the build-up of high pressure pockets of gas generated by bacterial activity. A mosaic of cracks over the icy surface of Europa (which has a diameter of 3,130 km) has led to speculations about the existence of a subsurface ocean that can harbour microbial life. These cracks are caused by tidal interactions with the planet Jupiter, which also serve to heat and maintain a warm liquid interior in Europa. Close-up views of the cracks were obtained by cameras on board the Galileo orbiter in 1997. Organic pigments seem to outline the cracks, some of which have dark soft edges, whilst others are interspersed with dark dots. This could imply that water laden with microorganisms oozed out of the cracks and refroze in relatively recent times. These discoveries give further credence both to the presence of a subsurface ocean on Europa, and to the possible existence of microbial life within it.

More recent Galileo discoveries relate to the largest of the Jovian satellites, Ganymede. At close quarters areas of Ganymede's icy surface show a network of ridges and cracks generally similar to Europa. It appears that ice tectonics and volcanism have operated on this satellite as well, possibly mediated by microbiology. Galileo spectrometers have also detected the presence of complex organic cyanides on Ganymede, from which a connection with life could be inferred.

A project of considerable future interest is the proposed *in situ* exploration of all three of Jupiter's icy moons, Callisto, Europa as well as Ganymede, the funding for which was approved by ESA (the European

Space Agency) in 2012. All three moons have tidally heated subsurface liquid oceans, and so are likely homes for life. Although this mission will not be launched for another decade and would reach Jupiter only in 2030, the prospect of a major astrobiology breakthrough at this time remains a strong possibility.

The Saturnian satellites are poorly understood at the present time, but prospects for microbial life are looking good. The Cassini–Huygens mission that was launched in October 1997 ejected a lander and probe to land on the surface of the 5,150 km-diameter Saturnian moon Titan in 2005. This mission has provided a wealth of information that has a bearing on the habitability of this satellite. Titan is the only natural satellite in the solar system that has a fully developed atmosphere which is chemically active and rich in molecular hydrogen and organic compounds. There is also tentative evidence for a layer of subsurface liquid water and ammonia, thereby enhancing prospects of life. In 2010, scientists analysing data from the Cassini–Huygens mission reported anomalies in Titan's atmosphere, including high methane content near the surface, which could be indicative of the presence of methane-producing organisms.

Chapter 10

Search for Exoplanets

The history of assertions about alien planets goes back a long time. The pre-Socratic philosopher Metrodorus of Chios (c. 400 BC) wrote: "It is unnatural in a large field to have only one shaft of wheat and in the infinite universe only one living world." The Roman poet Titus Lucretius Carus (c. 99–55 BC) stated: "Nothing in the universe is unique and alone, and therefore in other regions there must be other Earths inhabited by different tribes of men and breeds of beasts." In Indian and Asian philosophy the many worlds idea extends further back in time even into prehistory. The Vedas going back 3,000 years or more dwell on similar cosmological themes, and such ideas are encapsulated in Buddhist scriptures. In the comprehensive Buddhist Theravada text *Visuddhimagga* by Buddhaghosa written in c. 430 AD in Sri Lanka it is stated:

"... As far as these suns and moons revolve, shedding their light in space, so far extends the thousand-fold world system. In it there are a thousand suns, a thousand moons, a thousand inhabited Earths and a thousand heavenly bodies. This is called the thousand-fold minor world system..."

Here a multiplicity of inhabited earths and moons around other stars is clearly stated.

Assertions of a similar kind in Western Europe in the post-Classical period had to await the successful completion of the Copernican revolution. Giordano Bruno (1548–1600), an Italian monk, echoed closely the sentiments expressed in the *Visuddhimagga,* being prompted by new discoveries that were heralding a heliocentric world view. He went beyond the restricted Copernican model, however, by suggesting that stars were suns with their own planets orbiting around them, and furthermore that they were inhabited by alien beings.

"Innumerable suns exist; innumerable earths revolve around these suns in a manner similar to the way the seven planets revolve around our sun. Living beings inhabit these worlds." (*De l'Infinito Universo et Mondi*, Giordano Bruno, 1584)

Whilst the pronouncements of the *Visuddhimagga* fell within the cultural bounds of Hinduism and Buddhism, Bruno's were not compatible with the reigning Papal doctrine in Western Europe. For his impiety he was tried at the Inquisition and burnt to death.

The scientific search for alien planets had, however, to wait a full four and a half centuries after Bruno's death. Hubble telescope images first revealed the presence of many protoplanetary discs which show edge-on views of planetary systems as they were beginning to be formed. One such disc is seen edge-on around the star β Pictoris (Fig. 10.1).

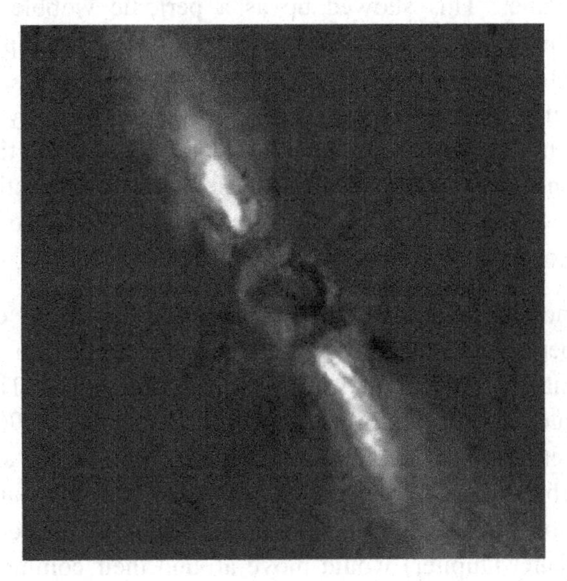

Fig. 10.1. Beta Pictoris showing an example of an edge-on view of a newly forming planetary system.

The detection of fully-fledged extrasolar planets telescopically stretches modern astronomical techniques to the limit. If a planetary system like ours existed around our nearest neighbour α Centauri, its Jupiter counterpart would lie only 1″ away from the central star, which would not be detectable in visible light against the brilliance of the central stellar disc even with the resolving power of our most powerful modern telescopes. Indirect methods of extrasolar planet detection are clearly required.

If a planet orbits a star, Newtonian dynamics tells us that both the planet and the star must each turn around their common centre of mass. In the case of a massive planet (like Jupiter) orbiting a Sun-like star, the centre of mass would in general lie outside the body of the planet on the line joining the planet and the star.

In April 1960 Peter van de Kamp claimed to have detected two massive planets orbiting the faint red dwarf Barnard's star some 6 light years away (van de Kamp, 1962). Van de Kamp's claims were based on observing the gravitational tugging effect on the star that was exerted by an unseen planet. This showed up as a periodic wobble of the star relative to the background of distant stars. Van de Kamp's claim of detecting such an effect was refuted in the mid-1970's on the basis that it arose from instrumental effects. However, the possibility of smaller planets orbiting Barnard's star could not be ruled out at the time. Van de Kamp's pioneering work and his basic idea of detecting "stellar wobbles" for finding unseen planets effectively gave rise to the modern science of exoplanet detection in the mid-1990's.

The method of looking for stellar wobbles has been extended to search for periodic Doppler shifts in the spectral lines of a star as it moved alternately towards and away from the observer. This Doppler shift technique was successfully deployed from the mid-1990's onwards for the detection of motions of central stars possessing Jupiter-mass planets in orbit. Viewed from the vantage of α Centauri our own Sun–Jupiter system would be detectable from such dynamical effects. The star (Sun) and planet (Jupiter) would move around their common centre of mass with the period of the planet (11.9 yr), and this effect will be visible as a small regular Doppler wobble in the star's apparent path in the sky.

Michel Mayor and Didier Queloz were the first to use the Doppler technique to discover many extrasolar systems, starting with the initial discovery of a Jupiter-mass planet around 51 Peg in the constellation of Pegasus 50 light years away (Mayor and Queloz, 1995).

This technique, however, has a strong selective bias to find Jupiter-mass planets in relatively close proximity to a central star. An exception to this rule has recently emerged with the deployment of an ultra-sensitive planet-hunting spectrograph HARPS on the ESO 3.6 metre telescope. Using this instrument several Neptune-sized and Earth-sized planets have so far been discovered. With continuing improvements in electronic sensors to record telescope images and the development of computer software to unravel minute fluctuations of starlight intensity, we are now on the threshold of a new era of planetary detections. The race is on to build better instruments and other techniques that can find more planets, especially those that are similar to our own in their capacity to support life.

Gravitational micro-lensing is another method of detecting extrasolar planets that has been used with some measure of success. In this process the gravitational field of a star–planet system acts like a lens, magnifying the light of a background distant star that lies in exactly the same direction. Due to the relative motion of the stars, lensing events tend to be brief, lasting only for days. Within the amplitude peak due to lensing by the star there is a brief spike if there exists an orbiting planet.

More than a thousand such events have been observed over the past ten years. If the foreground lensing star has a planet, then that planet's own gravitational field can make a detectable contribution to the lensing effect. Since that requires a highly improbable alignment, a very large number of distant stars must be continuously monitored in order to detect planetary micro-lensing contributions at a reasonable rate. This method is most fruitful for planets between Earth and the centre of the galaxy, as the galactic centre provides a large number of background stars.

A habitable zone around a star is defined as the range of radial distance in which a planet can maintain the conditions needed for life. This includes the requirement for liquid water at or near the surface, and ideally also a planet that can retain an atmosphere for timescales during which life can evolve. If the planet is too close to the star, surface temperatures would exceed the critical value for liquid water and if it is too far away the water will be in the form of solid ice. Another condition for a stable habitable planet is that it is not too close to a

Jupiter-sized planet whose interactions could lead to it being perturbed inwards or outwards (away from the habitable zone) on timescales that are too short.

Water will remain liquid under a pressure of 1 bar (terrestrial sea-level pressure) between 0°C and 100°C. If complicating factors, such as the effect of an atmospheric greenhouse are ignored, a habitable zone for Earth-type life could be defined simply as the distance from a star where the effective temperature falls in the range 273–373 K. In the case of the Sun this condition yields a range of radial distance for the habitable zone of 0.8–1.5 AU. The relevant range for other stars can now be calculated easily as $r \propto L^{1/2}$. Furthermore since for main sequence stars the luminosity of the star, $L \propto M^{3.5}$, where M is the mass, we have the simple result $r \propto M^{1.75}$. More realistic models of habitable zones around stars have been discussed by Franck *et al.* (2003) among others, taking account of factors such as an atmospheric greenhouse.

NASA's exoplanet mission, Kepler, was launched in 2009. This mission was designed to detect Earth-sized planets in the habitable zones of Sun-like stars by measuring minute fluctuations of intensity due to planets transiting in front. Within the first few months of its operation 5 new planets with sizes ranging from Jovian to Neptunian radii were detected (Borucki *et al.*, 2010). Upto November 2013 there have been a total of 2,740 candidate exoplanets, with a distribution of sizes/masses as shown in Fig. 10.2.

Amongst these detections are a population of relatively small planets that are looking more likely now to be exceedingly common in the galaxy. One of the more interesting discoveries is that of the 5-member planetary system Kepler62 some 1200 light years away. Two of these 5 planets have masses similar to the Earth and may well be capable of supporting life.

In general red dwarf stars with masses in the range 0.1–0.6 solar masses appear to offer the best scope for hosting habitable planets on a large scale. Although the luminosities of red dwarfs are typically only a tenth of a percent of the solar value (range ~0.01–3% solar luminosity), habitable exoplanets can still exist in orbits that lie close enough to parent stars. In such objects a combination of greenhouse gases and tidal locking could provide permanently warm life-friendly hemispheres in which biology could be expected to thrive. Ravi Kopparapu *et al.* (2013) and Dressing and Charbonneau (2013) have recently argued that M

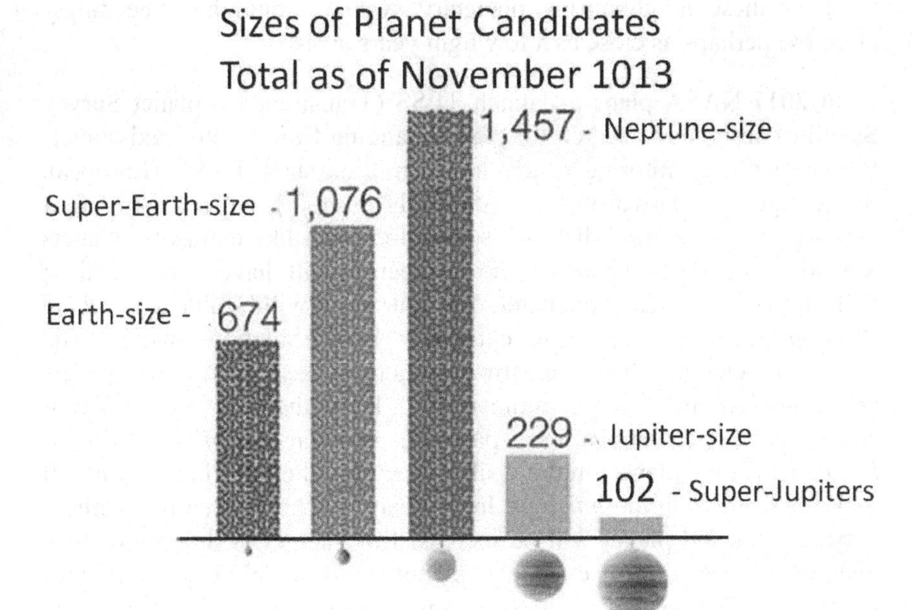

Fig. 10.2. Sizes of candidate exoplanets detected by Kepler mission up to November 2013. (Earth-size: <1.25 R_E; Super-Earth: 1.25–2 R_E; Neptune-size: 2–6 R_E; Jupiter-size: 6–15 R_E; Super-Jupiter: >15 R_E)

dwarf stars with surface temperatures in the range 2,600 K to 7,200 K have their habitable zone farther away from the stars than was previously thought. Applying their revised habitable zone criterion to red dwarfs pushes up the fraction of M dwarfs having habitable planets to about 50%.

Extrapolating from the restricted sampling of small planets observed around M dwarf stars by the Kepler spacecraft, these estimates indicate that there might be upwards of 100 billion Earth-sized planets lying within the habitable zones of M dwarf stars throughout our galaxy. And since the total population of these stars themselves is estimated to be around 100 billion in the Milky Way — we have an average of one Earth-type planet for every red dwarf star in our galaxy. Moreover since our solar system is surrounded by red dwarfs — very cool, very dim stars not visible to the naked eye (less than a thousandth the brightness of the

Sun) — these neighbouring planetary systems could be exceedingly close by, perhaps as close as a few light years away.

In 2017 NASA plans to launch TESS (Transiting Exoplanet Survey Satellite) designed to search for planets ranging from Earth-sized planets to gas giants, monitoring nearly half a million stars. ESA's (European Space Agency) Darwin mission (currently on hold) will consist of four "*free-flying spacecraft*" that will search for Earth-like extrasolar planets around other stars. Three of the spacecraft will have accurate light collectors of 3–4 meters diameter. The spacecraft will be able to *combine their mirror size* to provide extremely high-resolution images. The optical wavelengths at which astronomers have been looking for planets are *absorbed* by Earth's atmosphere. Thus there is a significant advantage of observations from space. Because stars are *billions of times brighter* than the planets in the visible spectrum, the Darwin mission will look for features in the *infrared.* In this way it is hoped that the contrast between star and planet will be increased by factors of thousands, thus making the detection of extrasolar planets considerably easier. Another advantage of observing at infrared wavelengths is that life on Earth is most easily detected through infrared. The Darwin observatories will be able to detect spectroscopically the same gases present in Earth's atmosphere which are signatures of life (oxygen, carbon dioxide and methane for example) if they exist in other extrasolar planets.

With confidence growing that planetary systems more or less like our own are exceedingly common, a generalised cosmological principle of mediocrity may be seen to be confirmed. If so we may soon find that habitable planets and 'solar systems' like our own are exceedingly commonplace. It follows that the most improbable event of life's origin *does not* need to be replicated independently in every planetary abode. According to panspermia theory the legacy of cosmic life will be out there in space, ever-ready to be taken up gratis by habitable comets, satellites or planets as and when they form.

Cometary panspermia involves the expulsion of individual microbes or small clumps of microbes from comets that seed recipient planets with life (Hoyle and Wickramasinghe, 2000). The first direct evidence that supported this hypothesis came on 31 March 1986 when Comet Halley made its last perihelion approach. The points in Fig. 4.4 (Chapter 4) represent observations made at the Anglo-Australian Telescope of infrared radiation emitted by a cloud of small particles in the vicinity of the nucleus of Comet Halley. The agreement between the bacterial model and the observations is striking, at least to the extent that we can say with certainty that the dust particles ejected from Comet Halley possessed infrared properties that are the same as those of bacteria. Infrared radiation from particles around Comet Halley was also measured on 3 March and 1 April 1986. On both these days the emission was weaker than on 31 March, implying that the cloud of particles — estimated to have a total mass of about a million tonnes — observed on 31 March were ejected from the comet during the previous 12 hours or so, and that by 1 April the clouds had largely spread out, dissipating into surrounding space. Comets would therefore seem to be capable of spewing out bacteria at a rate of about a million tonnes per day — about 10^{25} individual bacteria.

Such vast numbers of microorganisms expelled from comets can under suitable conditions reach other neighbouring planetary systems which we saw may be only a few light years away. The light from the Sun incident on a bacterium or clump of bacteria exerts a radially directed outward force P due to transfer of momentum, a photon of energy $h\nu$ carrying momentum $h\nu/c$. The star's gravitational attraction G acts in the opposite direction and both forces vary inversely with distance from the star. In situations where the ratio P/G exceeds unity, these particles are expelled entirely from the solar system. We can show that for a typical clump of bacteria emerging from a comet, expulsion occurs at a speed of some 30 km/s (Wickramasinghe *et al.*, 2010). With our estimated distance to the next nearest neighbour of 4–5 light years the time taken to cross this distance is only a few tens of thousands of years.

For such duration of exposure to low fluxes of low-energy cosmic rays in interstellar space, the loss of viability of a bacterium (such as *B. subtlis*) would be negligible (Horneck *et al.*, 2002; Lage *et al.*, 2012). An outer layer of carbonised material only 0.02 micron thick, which would form naturally due to exposure to space conditions, like sunscreen, protects the interior from ultraviolet light. With an estimated 10^{26} bacteria released from every Halley-like comet near perihelion the opportunities for some to infect and seed neighbouring planetary systems would be immense. From a single point source, say in the solar system, we can thus show that the entire galaxy could be infected in such a stepwise progression in less than 10 billion years, the age of low mass stars in the galactic disc.

In the above discussion we have dealt only with planets orbiting their parent stars. The possibility of very much larger numbers of free-floating planets unconnected to any star was first suggested by the pioneering work of R. Schild (1996), who measured effects of gravitational lensing of distant quasars due to intervening free-floating planet-sized bodies. Recently several groups of investigators have argued that a few billion such objects could exist in the galaxy (Cassan *et al.*, 2012; Sumi *et al.*, 2011).

In the model of the origin of life we described in an earlier chapter, life-bearing planets originated in the early Universe within a few million years of the Big Bang, and these planets make up most of the so-called "missing mass" of galaxies. We have recently calculated that such a planetary body would cross the inner solar system every 25 million years on the average, and during each transit zodiacal dust, including a component of the solar system's living cells, becomes implanted at the surface of the iterant planet. The free-floating planets would then have the added property of mixing the products of local biological evolution on a galaxy-wide scale (Wickramasinghe *et al.*, 2012).

Chapter 11

Search for Extraterrestrial Intelligence

The Earth is home to many billions of species of microorganisms, plants and animals. Single-celled microorganisms have evolved over 4 billion years to generate the marvellous spectacle of life we see today. Perched at the summit of this evolutionary pile is *Homo sapiens*. There are at present over six thousand million individual humans grouped into 221 separate nation states each possessing diverse histories. In the year 2014 an estimated 80 percent of the total human population live precariously on the verge of subsistence. The human population is further divided by religious and political ideologies that give rise to continuing competition and conflict over territory and resources. In this respect we are no different from lower life forms, the analogy remaining valid all the way down to colonies of bacteria and fungi.

A few million years ago our arboreal ancestors came down from the trees, began to walk on two legs and use their hands to make tools. These developments, combined with a growth of intelligence and intellectual capacity, contributed to our ability to harness the natural resources of the planet more efficiently than any rivals. Our development as a species over the past six thousand years can be seen as an expression of ever-increasing levels of what we recognise as "intelligence". Over the past few hundred years this process has accelerated at an unprecedented rate. Even in the fifty years our species has made enormous strides of technological progress including the exploration of space and unlocking the mysteries of the Universe. There is no reason to think that we may be anywhere near the end of the road in such pursuits. It is conceivable that the experiment of intelligence on the Earth has scarcely begun. Our quest to discover extraterrestrial intelligence must be seen against such a backdrop.

The search for extraterrestrial intelligence is based on three assumptions:

(a) such intelligence (ETI) exists;
(b) it is our duty as humans to attempt to make contact with such intelligence;
(c) the relevant technologies for such an endeavour — for sending signals, and receiving them — currently exist.

The programs currently in place to search for extraterrestrial intelligence are grouped under the acronym SETI. Those who believe in SETI take the three assumptions stated above for granted. But as would be expected, there are many who obstinately assert that *we are alone*, and so SETI would be meaningless and irrelevant. Modern astronomical studies discussed in the last chapter have given us every encouragement to believe that habitable planets like Earth are exceedingly common. We have also argued that the genetic components that were assembled into the complex tapestry of terrestrial life are also widely distributed throughout the cosmos. So any reservations that SETI sceptics may have must now turn on the argument that the emergence of intelligence from primitive life is an exceedingly rare or even a unique event in the Universe. Notwithstanding such doubts, progress in the search for extraterrestrial intelligence has continued from its earliest beginnings over a century ago.

As early as 1896, Nikola Tesla, a telephone engineer who worked with Thomas Edison, made the first known suggestion that radio could be used to contact extraterrestrial civilisations. The technical feasibility of detecting extraterrestrial intelligence by such means has been taken for granted for at least four decades. Philip Morrison and Giuseppe Cocconi first drew attention to the possibility of searching the microwave spectrum of cosmic sources, and suggested particular frequencies as well as a set of initial targets (Morrison and Cocconi, 1959). In 1960, Frank Drake conducted the first modern SETI experiment, named "Project Ozma", using the 26 metre radio telescope at Green Bank, West Virginia. The targets were two Sun-like stars tau Ceti and epsilon Eridani. Searches started over narrow wavebands centred around 21 cm, the famous neutral hydrogen line (the frequency at which hydrogen atoms absorb radio waves), but subsequently other wavelengths and multichannel detections have been attempted.

Historically two episodes of alleged ETI detection have been recorded in the past half century. One was dismissed as an error of judgement, the other still remains an unsolved mystery. In the winter of 1967 Jocelyn Bell and Anthony Hewish discovered a source of radio waves from a point source in the sky that was pulsing at the rate of once per minute. This source was later found to be a new type of astronomical object — a pulsar (PSR B1919 + 21), a rapidly rotating neutron star. For some weeks after its discovery, before its natural cause was understood, there were serious discussions of the possibility that this was indeed the sign of an extraterrestrial intelligence. Further implications of such a contingency were also under serious discussion albeit briefly. How do we verify and validate such a conclusion? How do we announce it, and could such a discovery be perceived as potentially dangerous for humanity?

The next more serious episode was one that has come to be nicknamed the "Wow!" signal. This was a brief burst of radio emission detected on August 15, 1977 by Jerry Ehman who was working at the time on a SETI project at the Big Ear Radio Telescope in Ohio, USA. The intensity of the signal coming from the direction of Sagittarius was observed to rise and fall over a period of 72 seconds. It was considered to be nearly impossible for any Earth-bound object to match the observed characteristics of the signal. If the signal did indeed originate in space, it was either a hitherto unknown astrophysical phenomenon (like in the case of pulsars) or it truly was an intercepted alien signal. The nearest star in the direction the telescope was pointing was some 220 light years away. The same star has been examined for over 3 decades, but the mystery signal was never repeated.

Nowadays, there is a growing trend to turn to the optical waveband, in the belief that laser signalling may have been used more advantageously by some of our more enlightened extraterrestrial neighbours. The idea of optical SETI was suggested by Schwartz and Townes and its deployment has so far not led to any result (Schwartz and Townes, 1961).

So much for feasibility and motivation, but let us briefly examine what the negative case might possibly be against SETI. In the 1950's Enrico Fermi stated that if technologically advanced civilisations are common in the universe they should be easily detectable, and indeed detected by now. Their silence or invisibility arguably poses a mystery and this simple fact may be used as an argument against SETI. This is the so-called Fermi paradox. A possible resolution of the paradox is to argue that while simple life (*e.g.* microorganisms) may well be abundant in the universe, intelligent life, or for that matter, even multicellular life may be very rare. In the view of the present author this assertion has no logical basis and might indeed run counter to available facts.

A superficial defence of the anti-SETI position could go as follows. Intelligence and the technology capable of SETI have arisen in only a span of a few thousand years out of a total history of terrestrial life that spans some 4,000 million years. Superficially, at least, the probability of such intelligence comes out at one in a million, and that would at best give only a million stars with inhabited planets carrying the requisite level of intelligence throughout the galaxy, at any given time. That may not be so bad after all, but some critics would go even further. They would say that intelligence on the Earth has arisen at the very end of a long series of multiple contingencies, a succession of random events, each having a vanishingly small probability. If you multiply a few thousand such infinitesimal probabilities you may end up with a chance of extraterrestrial intelligence (ETI) to be vanishingly small. On that argument Earth might be the only planet with intelligence in the entire Universe!

The ideas of panspermia which have been developed in this book offer a totally different landscape in which the logic of SETI can be restored. We argued that the emergence of the first life in the Universe has a probability so vanishingly small that it could reasonably have occurred once and only once in the entire history of the cosmos. The best setting for an origin would be for a spatially infinite universe, a universe that ranges far beyond the largest telescopes. Then the very small chance of obtaining a replicative primitive cell will bear fruit somewhere and, when it does, replication in a suitable astronomical setting will cause an enormous number of copies of the first cells to be produced. It is here that the immense replicative power of biology shows to great advantage.

According to this model no great innovation in biology ever happened on the Earth. The Earth is merely a receiving station where cosmically derived genes were assembled and life thereby evolved culminating in *Homo sapiens*. On this view of the origin of life there would be little variation in the forms to which the process gives rise, at least so far as basic genes are concerned, over the whole of our galaxy or even over all galaxies in the Universe. On other planets around other stars the same processes of assembly of cosmic genes as happened on Earth would also operate. Life would thus inevitably develop on every habitable planet assembled from the same all-pervasive cosmic genes. Intelligence — leading up to *at least* the level found in humans — is part and parcel of the package of cosmic evolution and should show up inevitably in the evolutionary history of life on every inhabited planet. It may come sooner rather than later depending on particular circumstances, and it could last a short time or a very long time, again depending on local contingencies. But like the emergence of life itself, the development of intelligence would also be a cosmic imperative. The simple logic is that evolution, which involves assembly of a cosmically derived blueprint, must converge everywhere to the same result. This is evident, for example, in the emergence of the eye that occurred independently on at least three occasions in the development of life on the Earth. So it must also be for intelligence on a cosmic scale.

In 1971, the US Space Agency NASA funded a SETI program led by Franck Drake. In recent years, however, institutional support for such studies has fizzled out due to funding constraints and competing priorities. A report proposing the construction of a radio telescope array with 1,500 dishes known as "Project Cyclops" has now gone through many iterations since the 1970's, but its full realisation is still only a distant dream. A somewhat more modest modern successor to Project Cyclops is the Allen Telescope Array (ATA) formerly known as the One Hectare Telescope, currently located at the Hat Creek Radio Observatory in northern California. The ATA is now a facility supported and managed by a privately funded organisation, the Stanford Research Institute. The ultimate plan is to have an array of 350 antennae scanning over a wide range of wavelengths and offering the best chance of detecting an intelligent signal. However, with just 42 antennae in operation since 2007, the project still limps along, being passed on from one public institution to another. The reluctance of any government institution or university to sustain an involvement is clear proof that this type of innovative science cannot be supported from public funds. Scientific authorities that decide such matters may also have a subconscious fear of actually making a discovery that may change the face of science forever.

It is clear, from the way this particular project has survived, that science will not easily be stifled by respect for authority. We can entertain an optimistic hope that with the advantages offered by

ATA — its wide field and instantaneous frequency coverage from 0.5–11.2 GHz — a confirmed "Wow!" signal will soon be received. This would surely be the most important development in the whole history of human civilisation.

Whilst radio SETI and optical SETI (OSETI) are the most favoured ways of searching for intelligence outside the Earth at the present time, an intriguing alternative can be described as biological SETI. Viral genomes, for instance, could carry coded information, if we can decipher them. The transfer of viruses across astronomical distances appears to be a fully feasible proposition. Thus intelligent messages could in principle be carried in viral genomes, and both bacteria and viruses could serve as transmitters of intelligent signals.

We mentioned earlier the proposal by Francis Crick and Leslie Orguel that the emergence of life on Earth was the result of directed panspermia — an artificially engineered bacterium or genome introduced to the Earth from outside (Crick and Orguel, 1973). This would demand the existence of an intelligence or super-intelligence capable of genetic engineering to the extent of creating the blueprint of carbon-based life artificially. Such an intelligent "working out" of the blueprint of life might at first glance be thought to be far-fetched, but is by no means impossible. In 2014 biochemists can perform feats of genetic engineering albeit in a limited way. Perhaps a few centuries from now human biochemists may be able to compute a genetic code and a set of crucial genes for life and disseminate them widely in the universe — in the form of viruses or bacteria for example. We have already seen that bacterial particles would have sizes that make them easily dispersed by radiation pressure of stars and could cross the few-light-year distance that has recently been estimated as the mean separation between habitable planets around M dwarf stars (Kopparapu *et al.*, 2013).

We discussed in earlier chapters that the functioning of biology depends crucially on a suite of enzymes (proteins) each of which is made up of a chain of amino acids. It is the arrangement of amino acids (of which there are 20 that are biologically relevant) within the chains that is crucial for life. And it is precisely this information that is held and transmitted by our genetic material DNA and mRNA by means of a genetic code.

This information of life in DNA or RNA is contained in the precise arrangements of four nitrogen-containing ring structures called bases. These are adenine (A), guanine (G), thymine (T) and cytosine (C) in DNA, but with uracil (U) replacing cytosine (C) in RNA. The coding is accomplished by arrangements of the bases, not singly but in triplets called codons. With few exceptions, each codon (triplet) encodes for one of the 20 amino acids. Thus, for example, the triplet UGC will always code to the amino acid cysteine and GCA will always code to glycine.

How this code originated has always been a mystery. In a recent paper published in the journal *Icarus*, shCherbak and Makukov (2013) have argued that an intelligent signal may be contained within our genetic code in the form of a mathematical and semantic message that cannot be accounted for by random Darwinian evolution. shCherbak and Makukov point out that the genetic code itself is the most invariant and durable construct across the whole of biology and once fixed the code will stay unchanged over cosmological timescales. The strict invariance of the genetic code therefore makes it a reliable mode of storage for an intelligent signal.

The claim is that statistical analysis of the human genome displays a thorough orderliness in the mapping between nucleotides and amino acids defined by the number of nucleons within them. The statement is that "Simple arrangements of the code reveal an ensemble of arithmetical and ideographical patterns of symbolic language." And this analysis is claimed to include the use of decimal notation, logical transformations, and use of the symbol *zero*. The hypothesis that all this arose by chance is rejected on the ground that it has an estimated probability of less than 10^{-13}. What the underlying message means, however, has not been discovered. But shCherbak and Makukov make the intriguing proposal that the genetic code could have been used to transmit a SETI-type message. This would not only be consistent with the ideas of panspermia discussed in earlier chapters, but it may signify the intervention of an intelligent agency that was perhaps responsible for originating the code itself.

A crucial quantity for assessing the merits of a SETI programme is the number of civilisations possessing the capacity to communicate (by radio or other means) that exists in the galaxy. Of the 100 billion habitable planetary systems that have been estimated to exist in the galaxy (see last chapter) let us suppose that 10% have planets like Earth on which intelligence can emerge. We can thus estimate 10 billion life-friendly planets in the galaxy. The now famous Drake equation tells us that if L is the average duration in years of an intelligence capable of SETI, we obtain an estimate of (L/10 billion years) × the number of planets as an estimate of the number of Earth-like planets with life at the required stage of advancement to engage in interstellar communication.

The total number therefore can be taken as being roughly equal to the number of years we expect such a civilisation to persist. From our human experience, our own capacity to engage in SETI is ~100 years. Already, at our stage of advancement of technology, we face an imminent risk of destroying ourselves as scientific technology advances, with primitive instincts of destructive competition remaining untamed. If a typical advanced civilisation inevitably destroys itself in 1,000 years, there will be only 1,000 civilisations with whom we can communicate in the galaxy. On the other hand if this time span is increased to 10 million years, then the number rises to 10 million. An enlightened civilisation on a planet like Earth could even survive and develop for upto 2 billion years, before an unavoidable stellar or planetary catastrophe intervenes. If so their numbers within the galaxy may well run into billions. There may be a case for saying that those civilisations that have overcome their primal tendencies of conflict, and developed enlightened pacifist philosophies dominate the cosmic scene through a process of natural selection.

It would be almost inevitable that a successful SETI contact — through radio, optical or biological means — would imply contact with a civilisation at a higher level of development than our own. If so there may be a great deal to learn from such contact. They may tell us how to avoid conflict. They may even tell us the nature of God if such exists, and thereby eliminate a major cause of strife and dissention amongst humans on our planet. We may even discover a cosmic religion in which petty squabbles such as we see on planet Earth will have no place!

Should we fear contact? The answer in the author's view is an emphatic NO. Our moment of first contact would be perhaps the most important, if not the most traumatic event, in the entire history of humanity. Mankind could instantly become enlightened to a degree that would scarcely be imagined. It would be like Neanderthals coming suddenly into contact with modern Men — our horizons would expand immeasurably.

Our genetic links to ET would, in accordance with ideas developed in this book, be similar to our links with past life forms on the Earth, for example to Neanderthals. ET would be made of the same cosmic genes and use the same genetic code. We humans appear to possess an uncanny inkling of this cosmic connection. It is perhaps no wonder that all our depictions and representations of ET in fiction and on the screen have not departed much from the body plans of creatures we know, and love or hate, on this planet. ET is conceived in the image of ourselves.

Chapter 12

Meteorite Clues

Meteorites were regarded as sacred objects and even worshipped by many ancient cultures. Meteorite falls witnessed as fireballs in the sky and often accompanied by sonic booms have always served to spark the human imagination.

In Greek mythology Phaethon was the young son of the Sun god Helios who begged his father to let him drive his chariot. Helios reluctantly allows his son to drive the chariot of the Sun, but Phaethon no sooner than he takes the reins loses control of the steeds. The Sun chariot veers out of control and crashes to Earth setting it on fire. Such myths are most likely a representation of major fireball events that occurred throughout history and prehistory. The tremendous power associated with such occurrences — fire, sound, smoke and destruction — would have evoked fear in the collective memory of ancient societies. It is no wonder then that the outcomes of such events, the meteorites, were often kept as sacred objects that symbolised divine power.

A giant iron meteorite that caused the 1.2 km-wide Meteor Crater in Arizona 50,000 years ago appears to have left vast amounts of meteorite fragments that were traded briskly, and these are widely distributed in archaeological finds in North America. Iron meteorites may have been especially treasured as raw material for knives and weapons before the discovery of iron smelting and the dawn of the Iron Age. Artefacts of this type have been recovered from the tombs of Egyptian Pharaohs. Close to the time of the Meteor Crater impact a similar meteorite or comet impact appears to have produced the 1.2 km-wide Lonar crater in India which has recently been dated at 52,000 ± 6,000 years. Many of the sacred stones (Shiva lingum) in Hindu temples dedicated to the god Shiva are believed to be derived from meteorites such as were probably associated with the Lonar crater.

The use of meteorites as sacred objects of worship was also widespread in Western Europe throughout historical times. It is evidenced in rocks that are preserved in temples such as the sacred stone in the temple of Apollo at Delphi, which is a rock that was said to have been thrown to Earth by Kronos the Greek god of time.

The connection of humanity with the external universe via comets, fireballs and meteorites continued to be expressed through myth and tradition well into classical times. The dawn of the Christian era, however, saw a reversal of this trend. The dominant belief thereafter was that the universe could not in any way be linked to the conduct human affairs. As we mentioned earlier, the execution of Giordano Bruno in 1600 and the trial of Galileo in 1633 began a trend of denial of unpalatable facts that related to our true place in the cosmos.

In the late 17th and early 18th centuries the French Academy of Sciences stubbornly denied all evidence for the fall of meteorites that were frequently witnessed by farmers. Similar denials were prevalent in the United States as well. For instance when Thomas Jefferson was informed that a meteorite fall in Connecticut in 1807 was seen by two Yale professors he is reported to have said "It is easier to believe that two Yankee professors would lie than that stones should fall from the skies." Unfortunately such denials became a bee dance-type ritual, and most scientific institutions and academies readily followed suit. Many precious collections of meteorites that were held in museums throughout Europe were destroyed, before the French Academy finally retracted from their denials and admitted that "stones can and do fall from the skies".

One of the earliest written records of a possible meteorite fall is to be found in a description in the *Bible* (Joshua 10:11), where a struggle between the Israelites and the Amorites is described thus:

"And it came to pass, as they fled from Israel, and were going down to Beth-Horon, that the Lord cast down great stones from Heaven upon them unto Azekah, and they died: there were more which died with hailstones than they whom the children of Israel slew with the sword."

Stones that fall to the Earth from space are of course "meteoroids". Speeds of incoming meteoroids range from about 72 km/s in a head-on collision with the Earth to about 13 km/s in an overtaking collision. If the meteoroids are large enough they would come through the atmosphere intact, with only an outer layer ablated by frictional heating; smaller ones burn out before reaching the surface.

The fate of incoming objects of arbitrary size depends on a variety of factors including mass, composition, porosity and angle of entry. The smallest meteoroids of sizes in the range millimetres to centimetres begin incandescing at heights of between 150 and 80 km above ground, the faster meteoroids becoming luminous at the greater heights. A meteoroid more massive than about 10 kg survives its journey through the atmosphere and lands intact as a meteorite. Only a relatively thin surface layer is heated and the interior of the object remains cold, so cold indeed that frost is known to have condensed on some meteorites when they were collected soon after a fall. Still larger meteoroids often break up into a multitude of fragments mid-air and produce showers of stones.

The smallest micron-sized meteoroids do not become seriously heated in transit through the atmosphere and drift down to settle at the Earth's surface. We saw in an earlier chapter that such particles often serve to nucleate raindrops, and that meteor showers and the discovery of freezing nuclei in clouds are correlated in time. Meteoroids of millimetre size burn out completely as shooting stars or meteors while they are still high in the atmosphere.

Meteors are seen in the night sky on a typical night with an average frequency of six to ten per hour. At certain well-defined times of the year when the Earth crosses the trails of cometary debris this average rate is greatly increased for several days. The meteors on such occasions appear to radiate from a point in the sky called the radiant, which defines the direction in which the Earth is moving through the particle stream. The position of the radiant amid the constellations gives the stream its name. The Taurids peaking in mid to late November have a radiant in the constellation of Taurus. Similarly the Leonids (November), Perseids (mid-August) and the Aquarids (July) are examples of regular meteor streams. Table 12.1 shows the data on these and other meteor streams.

Table 12.1. Regular meteor showers attributed to parent bodies.

Peak activity	Shower name	Average hourly count	Parent comet
January 3	Quadrantid	40	-
April 21	Lyrid	10	1861I (Thatcher)
May 4	Eta Aquarid	20	Halley
June 30	Beta Taurid	25	Encke
July 30	Delta Aquarid	20	-
August 11	Perseid	50	1862III (Swift Tuttle)
October 9	Draconid	Upto 500	Giacobini–Zinner
October 20	Orionid	30	Halley
November 7	Taurid	10	Encke
November 16	Leonid	12	1866I (Tuttle)
December 13	Geminid	50	3200 Phaeton?

In addition to meteoroids in cometary streams that give rise to the meteor showers of Table 12.1, the fracture and break-up of cometary crusts would give rise to larger fragments that end up as meteorites.

The case of Comet 2P/Encke and its associated Taurid complex is worthy of special note. This stream of debris that gives rise to regular Taurid meteor showers is thought to comprise of remnants of a giant comet that disintegrated some 20,000 to 30,000 years ago (Ascher and Clube, 1993). The resulting fragments ranging from 10μm-radius dust to fragments possibly larger than the Tunguska bolide (see Chapter 13) are distributed in the stream, being released both by normal cometary activity and occasionally by tidal interactions with planets. The Earth takes several weeks to pass through this rather wide stream and thus results in extended periods of meteor activity.

The larger meteoroids in the Taurid stream are thought to occupy an "inner" dense region that remains concentrated near the orbit of the stream's parent object. Ascher and Clube (1993) have argued that the orbits of meteoroids in this dense core have been subject to perturbations over thousands of years, and result in periodic episodes of intense bombardment of the Earth. We shall see later that this cometary meteor stream is a likely source of the meteoroids that resulted in a meteorite event in Sri Lanka in December 2012.

Meteoroids not necessarily associated with cometary meteor streams also fall to the Earth at a fairly steady rate. Over two thousand of these are estimated to fall every year, but only a few — less than a dozen or so — are tracked down and collected. The majority of these take place over the oceans or in remote areas and therefore go unnoticed. An average meteorite weighs several kilograms but very much larger falls sometime occur. The Allende meteorite that fell in Mexico in 1969 had a total mass of some 1,000 kilograms. The Murchison meteorite falling in the same year in Australia weighed about 225 kilograms.

Classification of Meteorites

The main purpose of classifying meteorites is to group specimens that share a common origin on a particular type of parent body. This could be a comet, asteroid, the Moon or a planet such as Mars, and within each meteorite class similarities of properties would be expected. Uniquely identifying a parent body is not always achievable because of insufficient information, but in general grouping according to chemical and mineralogical properties point to a common origin on a single type or class of parent body. The broadest classification scheme defines three main types:

Stony meteorites composed of between 70 and 90 percent silicate minerals, with smaller amounts of nickel and iron sulphide. These meteorites account for over 90 percent of recovered meteorite falls.

Iron meteorites mainly comprised of a nickel-iron alloy account for some 5 percent of the observed falls, but they are more easily recognised, and as was stated earlier served as sources of metallic iron for making implements in prehistoric times. Iron meteorites most probably originate in the cores of large asteroids or protoplanetary bodies, and are composed mainly of a nickel-iron alloy.

Stony-iron meteorites contain equal amounts of silicates and iron-nickel alloy, and also most probably originate in asteroids.

Unclassified meteorites represent perhaps largest class of meteorite with stones that fall from the sky that do not readily fit into any of the above categories. Our special interest in this category is due to the possibility that perhaps the most compelling evidence for extraterrestrial life may be contained within them. The Sri Lankan event of December

2012 possibly linked to the Taurid meteor shower may have led to meteorites in this category. Before discussing this particular event we shall describe studies over several decades that relate to carbonaceous chondrites.

Carbonaceous chondrites contain a few percent carbon in the form of organic material, hence its name carbonaceous. In a recently adopted classification scheme of carbonaceous chondrites, 8 main groups are defined by their best known prototypes, of which Ivuna and Mighei are among the most important. The Ivuna meteorite fell on December 16, 1938 in Ivuna, Tanzania. The Mighei meteorite fell in Ukraine in 1889 is generally considered to be unique amongst carbonaceous chondrites.

The CM chondrites are named after the type specimen Mighei and they contain water and complex organic compounds, including aromatic molecules, nucleobases and amino acids. Amongst the other carbonaceous meteorites that have been extensively studied in recent years are: the Orgueil meteorite that fell on May 14, 1864, near the town of Orgueil in southern France; the Allende meteorite that fell in Mexico on February 8, 1969; and the Murchison meteorite that fell near the town of Murchison, Victoria in Australia on 28 September 1969. The numerical designations of 1 and 2, *e.g.* CI1 and CM2 indicate the degree of water processing or aqueous alteration (alteration due to prolonged contact with water). This alteration occurred in the parent body — a presumed comet — under very low temperature conditions, probably in the temperature range of 20 to 50°C in a water-rich environment.

Radioactive dating techniques give an age of formation of carbonaceous chondrites of between 4.5 and 4.7 billion years, so that they are certainly older than the Earth's crust. There can be little doubt that these objects have undergone the gentlest possible thermal history since the time of their compaction from separate grains and molecules. They have not been heated at any time in their history to a temperature in excess of ~500 degrees Kelvin. Any heating above 500 K would have altered their primordial chemical and mineral composition that is preserved today.

Studies of particles extracted from carbonaceous chondrites have shown the presence of micron-sized clumps consisting of many separate grains each with a radius of about 100 Angstroms. There is also evidence that a minor component of the material in this meteorite class is genuinely extrasolar, that is to say, condensed into solid particles from a gaseous state in distant parts of the galaxy. This conclusion is drawn from the detection of several isotope ratios which are distinctly anomalous in relation to the solar system. One such discriminant is the ratio of neon isotopes ($^{20}Ne/^{22}Ne$) in embedded "presolar dust grains". The observed isotopic anomalies relative to solar system values are explained if the grains are condensed in the vicinity of a supernova. It can be shown that such condensation does indeed occur within the gaseous matter which flows outwards in an expanding envelope of a nova or supernova. These presolar grains in meteorites did not subsequently re-melt and still carry the imprints of "extinct radioactivity" derived from the source of their original condensation. We therefore have a tangible solid particle component of interstellar space arriving intact within these meteorites, thus connecting our solar system to distant parts of the galaxy.

Organics in Meteorites

The amazingly rich diversity of extraterrestrial organics recently discovered in the Murchison meteorite (Schmitt-Koplin *et al.*, 2010) comes as no surprise in the context of the ideas discussed in the earlier chapters. If cometary bodies are the carriers of microbial life, a diversity of organic molecules as rich, or possibly richer, than that which prevails naturally on the Earth is to be expected. Moreover, such molecules would have been generated overwhelmingly through biochemical rather than abiotic or prebiotic processes. Since a diverse extraterrestrial biology carried in comets would be expected to lead to a far greater range of possibilities than the limited subset of life selected by niches on the Earth, a higher level of diversity in the meteoritic organics is not at all surprising. It may not be necessary to assign such a diversity to prebiotic processes as suggested by Schmitt-Koplin. For instance "non-biological" amino acids — *e.g.* AIB and isovaline — that might be relevant to an exotic or alien biology were also found at the K–T boundary during the time of the extinction of the dinosaurs, and it can be argued that this material may have been deposited onto the Earth by comets. Similarly, the range of molecules discovered in the Murchison meteorite could include degradation products of a wide range of alternative non-terrestrial biologies.

Primordial bodies of the type discussed in an earlier chapter, endowed with radioactive heat sources, provide an ideal venue not only for an ultimate origin of life but also for the replication of microbial life (Gibson *et al.*, 2010). Comets are formed when such bodies become included in the material from which planets form. Within an individual comet endowed with nutrients and chemical energy a pre-existing microbiota can proliferate on a very short timescale. Thereafter an amplified microbial population becomes locked in a freeze-dried state until the comet comes to be peeled away layer by layer, thus releasing viable microbes into space. The expulsion of volatiles from a comet over many perihelion passages would eventually lead to a compaction of mineral grains in comets along with residual microorganisms that effectively "sediment" and fossilise. On this basis it is possible to understand an origin of carbonaceous chondrites such as the Murchison and Orgueil meteorite as fragments of "extinct" comets.

Organised Elements in Meteorites

The few percent carbon content of CM and CI carbonaceous chondrites, which occurs in the form of organic compounds, may be separated into two groups: those that are soluble in organic and inorganic solvents, and those that are not. It is the insoluble component that has proved more intractable and sparked off controversy. An intriguing debate followed a first report by G. Claus and B. Nagy (1961) and H. Urey (1966) in which they claimed to have discovered structures resembling microbial fossils in the Orgueil and Ivuna meteorites. Both these meteorites were actually seen to fall, tracked down and recovered. The Orgueil meteorite fell in France in 1864 and the Ivuna meteorite in central Africa in 1938.

Although electron micrographs revealed evidence of structures resembling cell walls and flagella and a wide range of biological morphologies, all these identifications came to be discredited because contamination by ragweed pollen was demonstrated in at least one instance. It became clear that much more careful attention to detail as well as technique was needed before this matter could advance any further. However, the lack of compelling evidence to exclude contamination led many scientists to become highly sceptical of the fossil explanation of the structures found, and alternative explanations were promptly put forward. One such alternative explanation was that the fossil-like structures (that were not contaminants) are mineral grains that had acquired coatings of organic molecules by some abiotic process.

Whilst a bee dance-type consensus against the biological nature of the "organised elements" or "microfossils" in meteorites soon came to be established, the matter was not really resolved in terms of science itself. Critics made their case for contamination so stridently that the scientific world became convinced that any claim of microfossils must be wrong. Claus apparently reneged under pressure, and Nagy retreated while continuing to hint in his writings that it might be so, rather in the style of Galileo's quietly whispered "*E pur si muove*" — and yet it moves!

Nearly two decades later the problem of microbial fossils in carbonaceous meteorites came to be re-examined by Hans D. Pflug with special attention being paid to sample preparation so as to avoid the criticism of earlier work (Pflug, 1984). Pflug used state-of-the-art equipment to prepare ultra-thin sections (<1 mm) of the Murchison meteorite in a contaminant free environment. The results are presented in Figs. 12.1 and 12.2.

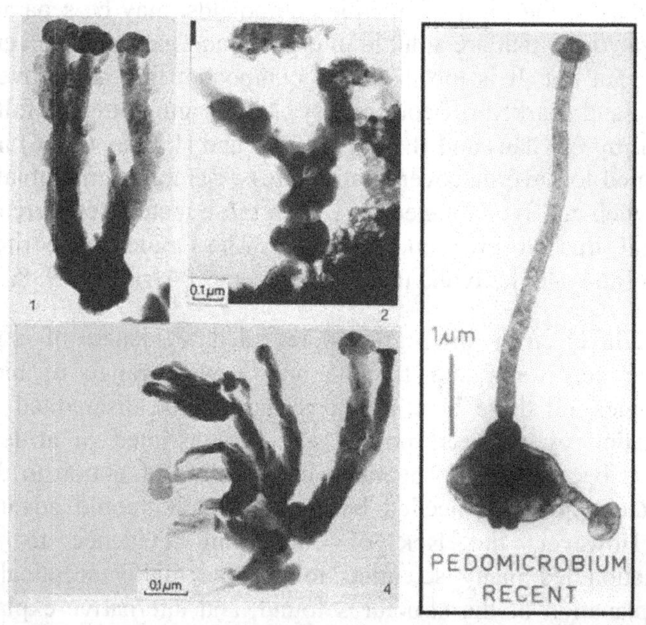

Fig. 12.1. The comparison of a characteristically biological structure in the Murchison meteorite with a similar structure corresponding to a modern iron-oxidising microorganism — *Pedomicrobium.*

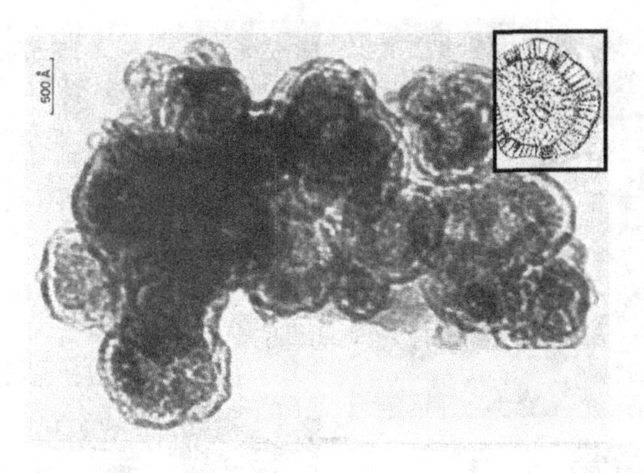

Fig. 12.2. An electron micrograph of a structure resembling a clump of viruses — influenza virus — also found in the Murchison meteorite. The drawing in the inset is a representation of a modern influenza virus displaying astounding similarities in structure to the putative clump of fossil viruses.

Further work by Pflug and Heinz (1997) confirmed these results and the criticism of contamination levelled against Claus and Nagy now became largely irrelevant.

Pflug's pioneering work was confirmed in large measure by Richard Hoover and his collaborators from 2000 to the present day (Hoover, 2005; Hoover, 2011). Hoover *et al.* showed that diverse microbial morphologies were present on freshly fractured surfaces of carbonaceous chondrites, including the Murchison and Tagish Lake meteorites. They consider various criteria for separating recent terrestrial contamination from indigenous microfossils and have diligently compiled an atlas of prospective bacterial fossils. Backscatter electron images at high resolution in Fig. 12.3 shows a particularly impressive example comparing indigenous structures on a freshly cleaved surface of the Murchison meteorite with living cyanobacteria. The oft-stated criticism of contamination is used to dismiss these identifications, but this is now very difficult to defend on a purely objective basis.

Fig. 12.3. Structures in the Murchison meteorite (Hoover, 2005) compared with living cyanobacteria.

Microfossils in Cometary Dust

On a much smaller scale than meteorites, clumps of interplanetary dust particles of cometary origin — micrometeorites — have been collected over many years by using sticky paper flown on U2 aircraft (Brownlee *et al*., 1977). These so-called Brownlee particles, which were mostly in the form of fluffy aggregates of siliceous dust, have also been found to contain extraterrestrial organic molecules, with a complexity and diversity approaching that recently reported for the Murchison meteorite (Clemett *et al*., 1993). In a few instances microbial morphologies were discovered within individual particles (Hoyle *et al*., 1985).

We have already referred to the isolation of culturable as well as viable but non-culturable microorganisms from cryosampler collections of stratospheric aerosols at a height of 41 km (Harris *et al.*, 2002; Narlikar *et al.*, 2003; Wainwright *et al.*, 2003; Shivaji *et al.*, 2010). As in the case of the Brownlee particles, aerosols obtained from the stratospheric samples were found to contain a rich harvest of pristine carbonaceous cometary dust. Scanning electron microscope studies combined with chemical element identifications revealed putative bio-fossils — organic-walled hollow spheres about 10 μm across, similar to those found in carbonaceous chondrites as well as in ancient terrestrial sedimentary rocks and termed 'acritarchs' and cylindrical structures resembling diatoms. Examples of such structures are shown in Fig. 12.4 (Miyake *et al.*, 2010).

The structures in the lower frames in Fig. 12.4 resemble siliceous fragments of diatoms — sub-micron whiskers up to 10–15 μm long fibres of 1–2 μm diameter, occurring both singly and in complexes. We

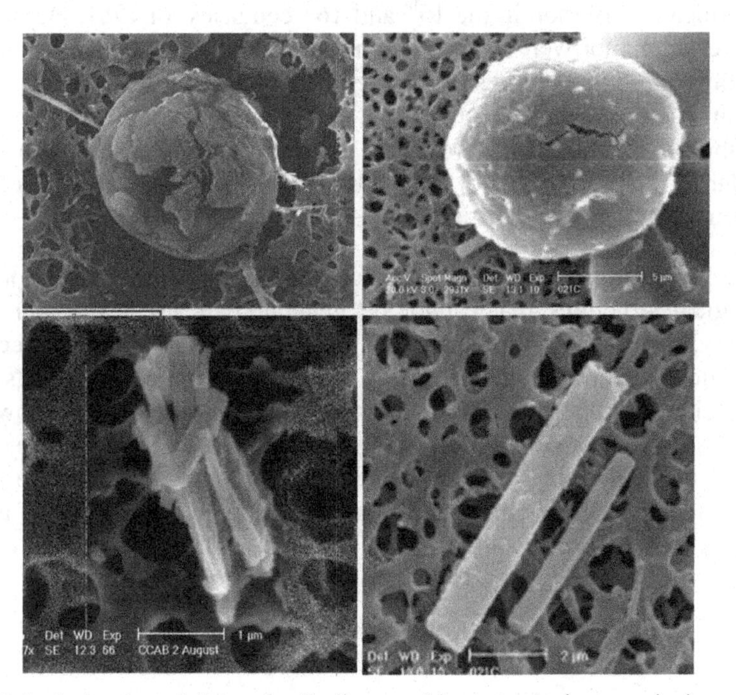

Fig. 12.4. Acritarchs and diatom fossils discovered in stratospheric atmospheric aerosols collected from 41 km in the Earth's atmosphere (Miyake *et al.*, 2010).

noted in an earlier chapter that in a more recent study led by Milton Wainwright, clumps of cometary dust and individual biological entities measuring 30–100 μm in radius were collected directly onto electron microscope stubs that were lofted by balloon to 27 km in the stratosphere. We already noted the discovery of unusual biological entities associated with particles that are so large that they could not have been lofted from the ground in any freak meteorological event. Impacts at speed of several cm/s have in fact left microcraters testifying to their space origin (see Fig. 8.5).

Unexpected Evidence from a New Class of Meteorite

Just as in the case of the Copernican revolution of the 16th century correct ideas in science tend to reaffirm their validity against the tide of prejudice repeatedly and in diverse and unexpected ways. Ultimately a paradigm shift is forced into place. Resistance to the theory of cometary panspermia has been every bit as ferocious as was the opposition to the Copernican revolution in the 15th and 16th centuries. In 1981, Hoyle and the present author were left in no doubt that Hans D. Pflug's evidence of microbial life in meteorites constituted decisive proof of the existence of life in comets. Feeble excuses that continue to be voiced about the possibility of inorganic processes producing filamentary structures similar to the cyanobacterial filaments in Fig. 12.3 have little or nothing in their favour.

If, however, a meteorite turned up that contained morphologies of biological structures that cannot be generated in any way other than biology, then victory for panspermia might at last have to be conceded. Biological structures that had for long been attributed to comets and interstellar dust are the class of photosynthetic microorganisms known as diatoms. These organisms are known to be prolific in the lakes and oceans of the Earth, including in the polar regions. So we have argued that they could also find a natural habitat in the subsurface liquid pools such as exist in comets and in satellites such as Europa (Hoover *et al.*, 1984).

Fig. 12.5. A selection of terrestrial diatom frustules.

It is estimated that there are over 100,000 distinct species of extant diatoms — marine and freshwater — each with distinctively sculptured siliceous outer coverings called frustules. These siliceous frustules are unmistakably generated biologically and their existence anywhere must constitute clear and unequivocal proof of life. Some examples of these structures are shown in Fig. 12.5.

It is a curious fact that has a bearing on our present discussion that diatoms appear very suddenly in the geological record some 180 million years ago. Considering the extreme stability of their silica shells, the absence of fossils in earlier epochs supports the idea of a cometary injection at a definite moment of geological time.

Polonnaruwa Meteorite

Serendipity intervened at this stage in favour of cometary panspermia — amazingly in the island of Sri Lanka that was known to Marco Polo as Serendip. Minutes after a large fireball was seen by a number of people in the skies over central Sri Lanka on 29 December 2012, a large meteorite disintegrated and fell in the village of Aralaganwila which is located a few miles away from the historic ancient city of Polonnaruwa.

We shall refer to these stones as "Polonnaruwa meteorites" even though the meteoritic community is still to accept this nomenclature. Figure 12.6 shows a specimen of this meteorite as well as the location in Sri Lanka where the fall occurred. At the time of entry into the Earth's atmosphere the parent body of this Polonnaruwa meteorite would have had most of its interior porous volume filled with water, volatile organics and living cells as we shall see. Strange odours were reported and a few people had their hands burnt in touching the rocks as they fell.

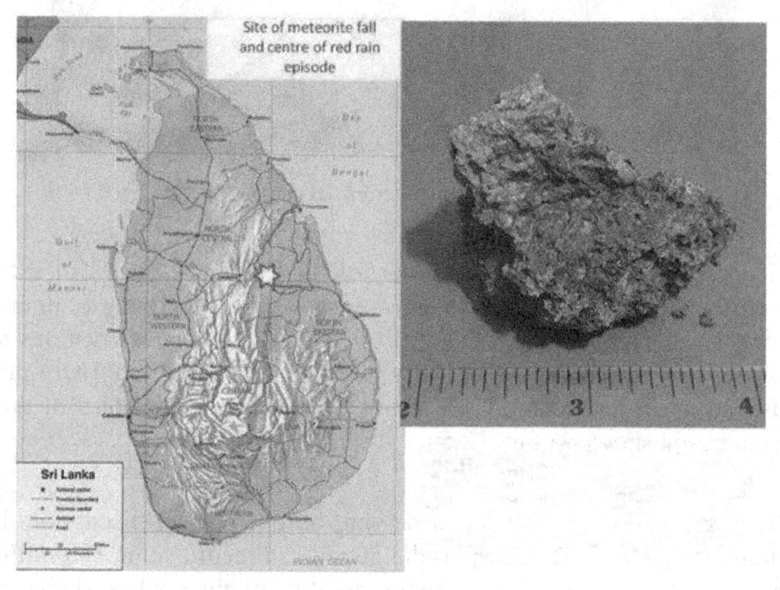

Fig. 12.6. Location of the Polonnaruwa meteorite fall and a specimen of stone.

The meteorite itself falls squarely in the category of "unclassified" and causes consternation amongst meteoriticists. However, there is no doubt that it fell from the sky and also that it is linked with an extended Taurid stream, indicated by the location in the constellation of Taurus from which the fireball emerged. The stones are highly porous (80% porosity) and are rich in Si and K, with a carbon content of a few percent. Detailed physical, chemical and mineralogical analysis of the Polonnaruwa meteorite has been conducted by Jamie Wallis (Wallis, 2014) and there is little doubt that we have here a new class of meteorite, representing the siliceous residue of a cometary lake of the type we discussed in Chapter 8. It has an amorphous SiO_2 melt matrix within which are included 10 micron-sized grains comprised of high impact minerals — anorthoclase, albite and anorthite, bearing testimony to an impact ejection event. It is also found to have a very low bound H_2O fraction and the carbonaceous component of the stones shows strong nitrogen depletion with N/C values less than 0.3%.

Fragments of freshly cleaved interior surfaces of these meteorites were mounted on aluminium stubs and examined in a scanning electron microscope (Wickramasinghe *et al.*, 2013a, b; Wallis *et al.*, 2013a, b). Images of the sample showed a wide range of distinctly biological structures that were distributed and enmeshed within the rock matrix, examples of which are shown in Figs. 12.7 and 12.8.

The structures in the top row of Fig. 12.7 resemble extinct microbial fossil known as an acritarch, and the images in the bottom row and also in Fig. 12.8 correspond to species of diatoms with distinct and

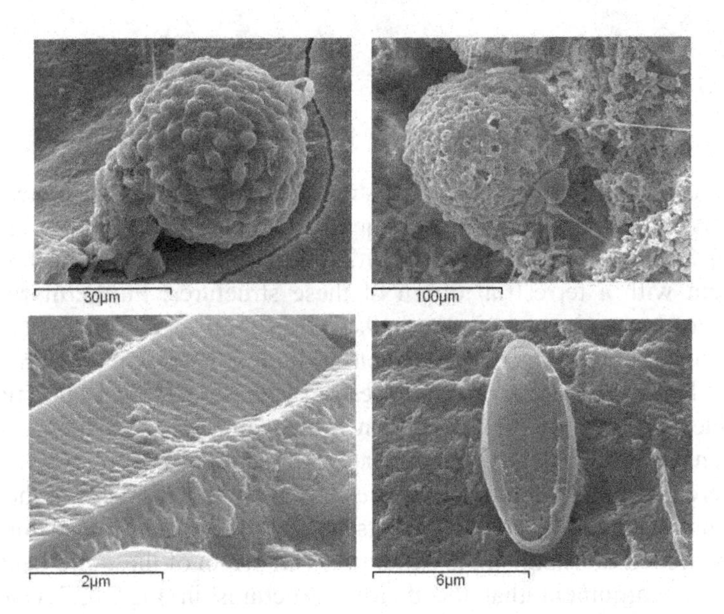

Fig. 12.7. Fossilised acritarch and diatoms in the Polonnaruwa meteorite.

Fig. 12.8. Fossilised diatoms in the Polonnaruwa meteorite.

unambiguous morphologies. These structures cannot be interpreted on the basis of contamination, or mineralogical artefacts. Wallis *et al.* (2013b) have shown that the disposition of oxygen isotopes cannot be consistent with a terrestrial origin of these structures. Furthermore the analysis of trace elements in the bulk composition of the meteorite has yielded high abundances of the element iridium at levels of 7–8 parts per million. This is in great excess of terrestrial values — about 10,000 times the values appropriate for material in the oceanic crust — but is fully consistent with cometary and meteoric values. Indeed we noted elsewhere that high iridium abundances in terrestrial sediments at the K–T boundary pointed to comet impacts 65 million years ago. No further proof is needed to establish the extraterrestrial origin of these life-bearing stones. The argument that the diatom structures in Fig. 12.7 can be modern contaminants can be disposed on the basis of the low N/C ratio of the carbonaceous content of the meteorite that can only be understood if the biological structures are fossilised.

Chapter 13

Comet Impacts and Civilisation

In earlier chapters we argued that comets are of crucial importance in relation to the origin and evolution of life on our planet. In this chapter we consider the physical effects of comets and fragments of comets colliding with the Earth, and the consequences of such events in relation to life at a higher organisational level than hitherto considered — to the evolution of culture and religion.

We have already noticed that comets pursue elliptical orbits around the Sun, and that a 100 billion or so cometary bodies — objects left over from the formation of Uranus and Neptune — still surround the planetary system in the form of a gigantic shell of comets. Between the orbit of Neptune and the Oort cloud are several distinct families of comets, including the Kuiper belt of comets that pursue orbits in a plane inclined to the plane of the planets. There are also more scattered populations of "transneptunian objects", of which Pluto is one.

The effect of passing stars on the Oort cloud is to deflect individual comets and make them plunge in highly elongated elliptical orbits into the inner regions of the solar system. A comet's first incursion into the inner regions of the solar system takes place along an orbit that has a period in the range of hundreds of thousands of years or even longer. After a number of transits through the inner regions of the solar system a long-period comet becomes eventually transformed into one with a much shorter period due to the build-up of gravitational interactions with the giant planets Jupiter and Saturn. The shortest-period comets have periods of three to four years.

A class of comet that has recently come to the fore is that of the so-called giant comet with a typical radius in the range 30–200 kilometres.

The existence of such giant comets was first predicted by Victor Clube and Bill Napier (1990) but remained only as a conjecture until the discovery of the object known as Chiron in 1977. At first it appeared that Chiron may be just another asteroid moving in a peculiar orbit that lay between the orbits of Saturn and Uranus. In 1989, however, Chiron's cometary character was revealed when it was found to brighten enormously and develop an extended cometary coma. Chiron was the first giant comet to be discovered with a radius of some 115 kilometres. Subsequently a large number of similar objects have been found, defining a new class called the "Centaurs" mostly occupying the outer regions of the solar system beyond the orbit of Neptune.

A member of the class of giant comets to come to our attention in the late 1990's was Comet Hale–Bopp. This comet had an estimated diameter of 40 km and came towards the lower end of the size range expected for giant comets. Hale–Bopp made its closest approach to the Sun on 1 April 1997, and it is due to return again to the inner solar system in the year AD 4377. The effect of Jupiter on comets is markedly evident in the case of Comet Hale–Bopp. The orbital period as it came in from the depths of the solar system, was 4,200 years, but its passage past Jupiter at a distance of 65 million miles reduced this period to 2,380 years.

The first half billion years of the Earth's history was riddled with comet impacts. The collisions at this epoch essentially represented the final stages of the formation of the Earth. Although this initial impact-dominated epoch, known as the Hadean epoch, came to an end about 4 billion years ago, collisions with comets and fragments of comets did not come to a complete halt.

So what were the direct physical effects that resulted from the collision of comets and cometary fragments with our planet? Many of the planets and satellites of our solar system show heavily cratered surfaces. As a general rule planets with atmospheres exhibit less in the way of surviving scars, for the reason that impact scars are relatively quickly smoothed over by surface weathering. A close inspection of the Earth shows abundant evidence of impacts. There is now little doubt that the extinction of the dinosaurs and of over 75 percent of all genera of plant, animal and microbial life occurred 65 million years ago was a result of a comet impact. A process of this kind was first suggested by Hoyle and myself as early as 1978 (Hoyle and Wickramasinghe, 1978). In our paper we wrote that "Although more dramatic and seemingly more devastating, a direct hit would not necessarily be as far-reaching in its effects as the addition of 10^{14} grams of small grains... would be". Such a protracted episode of atmospheric "dusting" by comets, which we estimate would occur once in about 100 million years, would be by far the most important process in precipitating a major ecological catastrophe such as occurred around 65 million years ago. There is some evidence to suggest that the extinctions of species at this time occurred over a protracted period of some 100,000 years, peaking in a single crater-producing event at the midpoint of this interval. A veil of dust in the stratosphere would have the effect of dimming sunlight to a level that photosynthesis by microscopic plankton in the ocean at the bottom of the food chain of all animals would fall to a very low rate. Leaves would wither from trees, leading eventually to extinctions of large browsing animals, as happened in the case of the dinosaurs 65 million years ago.

A very similar process connecting the impacts of comets with the extinctions of species was being discussed in parallel by Bill Napier and Victor Clube, and published in a paper that appeared in *Nature* (Vol. 282, 29 November 1979). Two years after our comet impact theory was published, L.W. Alvarez, W. Alvarez, F. Asaro and H.V. Michel discovered an enhancement of the chemical element iridium in the Earth's sedimentary deposits of 65 million years ago, pointing clearly to the involvement of comets (Alvarez *et al.*, 1980). Iridium is exceedingly rare on the Earth but is much more abundant in comets and meteorites. Thus, whenever we see a layer of terrestrial sediment that is rich in iridium, we can be sure that a comet or comets collided with the Earth. Later studies showed that certain amino acids normally found in meteorites, but not so commonly in life, are also present in the same sediments, again pointing to the addition of cometary or meteoric material. A while later it was found that the comet collision that coincided with the extinction of the dinosaurs has also left its mark in the form of a crater buried in the seabed near the village of Chicxulub in the Yucatan peninsula.

Other large-scale extinctions of species at earlier times have also been recorded and found to be associated with enhancements of iridium in geological sediments, so a cometary connection can be inferred. Significant peaks in the extinctions of species have been discovered in the geological record at approximately 1.6, 11, 37, 66, 91, 113, 144, 176, 193, 216, 245 and 367 million years ago. Indeed, the entire pattern of mass extinctions of species over the past 400 million years is strongly suggestive of recurrent catastrophic events of external origin with a significant periodicity of about 26 million years.

We should not be content to think that dramatic effects of comet collisions are only confined to distant geological epochs. More recent episodes of comet and asteroid impacts have surely left their mark, not only as scars on the planet, but upon our history as we shall now show. To understand this process it is important to recall the nature of comets. Recent studies have shown that comets possess relatively fragile internal structures that can be easily disrupted.

On 25 July 2000, Comet Linear was heading towards its perihelion or closest approach to the Sun. It was in itself a relatively unimpressive object, visible only with a pair of binoculars, until it suddenly flared up and brightened, producing a huge coma and tail. Then it mysteriously disappeared for several days, but reappeared again as a much fainter comet. Long-exposure photographs revealed that the comet was split into over 20 separate small pieces with each piece contributing to a much weakened cometary coma. The weakly-welded multi-cracked structure of the comet was obviously torn apart by the force of the evaporated gas that built up huge pressures inside. The most recent long-period comet ISON suffered a similar fate as it approached perihelion on 28 November 2013. In this case no large surviving pieces appear to have emerged post perihelion, and the comet was probably reduced to debris stream of gas and dust.

Comets for the most part stay very far from the range of the inner planets occupying orbits that envelop the planetary system at a great distance. Occasionally a comet swings past the giant planet Jupiter. When this happens it is not only deflected in its orbit, so that it no longer returns to the most distant regions whence it came, it can also be shredded into pieces by the tidal forces exerted by Jupiter. This happened in the summer of 1992 for the now famous Comet Shoemaker–Levy 9. The comet broke up into a string of some 21 pieces. In this case the pieces were all pulled in to become satellites of Jupiter moving in orbits that slowly spiralled in towards the planet. Over the course of a single week in July 1994 all 21 pieces of Comet Shoemaker–Levy 9 crashed on to Jupiter.

The history of civilisation, if it is correctly deciphered, bears witness to the most recent chapter of collisions with comets and fragments of comets, collisions that in effect controlled the fate of mankind. The ideas that shall be described in the rest of this chapter follow generally the theories advanced by Napier and Clube but with modifications that were introduced by Hoyle and the present writer. Whilst Napier and Clube have provisionally identified the original giant comet as a progenitor of Comet Encke, with its stream of debris connected with the Taurid meteor stream, we have left unspecified the precise identity of the giant comet and so also its original orbit and associated meteor stream. The original giant comet itself would have long since ceased to exist, because it fragmented in the manner of Comet Linear or Comet Shoemaker–Levy, forming a commoving meteor stream comprised of billions of individual pieces.

The fragmenting giant comet, call it Comet X, may be estimated to have weighed of the order of 10^{16} tonnes, 1,000–10,000 times the mass of Halley's Comet, and measured some 200 kilometres across. It would have become perturbed by Jupiter, say 20,000 years ago, into a periodic orbit that crossed the Earth's orbit. Within a few orbital revolutions around the Sun fragmentation of the comet would have proceeded sequentially in the style of Comet Linear to produce disintegration into fragments of decreasing sizes, ranging from say 10 kilometres to 100 metres or less. On a ball-park estimate, the kilometre-sized pieces could number millions and the 100 metre-sized pieces could number billions. All these pieces would remain bunched together within a meteor stream and the evolution of their orbits will be such that the Earth encounters the debris stream with some well-defined periodicity.

The average time interval between episodes of bunched collisions with debris is difficult to pin down to any degree of precision. Encounters could start with the initial period of Comet X, but as the meteor stream disperses and evolves the situation will become more chaotic. We shall now proceed to chart the progress of Comet X and its progeny through time by attempting to identify crucial events in the history of our civilisation that appear to be connected with possible cometary incursions. In this way we avoid an awkward dynamical problem and work backwards from historical facts to determine the time between successive encounters with the meteor stream of Comet X.

Each encounter will be characterised by an epoch of bunched collisions with cometary missiles. During a bunched collision epoch, impacts of objects a few hundred metres across would pose a far greater threat than impacts of much larger pieces that would occur far less frequently. Interaction with smaller fragments would have occurred with menacing frequency during the bunched collision epochs, each of which may have lasted several decades.

During the past 20,000 years perhaps the most important geological event was the emergence of the Earth from the last ice age. This event may well have been triggered by impacts with cometary fragments. The process of unlocking the planet from its glacial state may have been accomplished in several stages, as indicated by the record of temperature from Greenland ice-core data as depicted in Fig. 13.1.

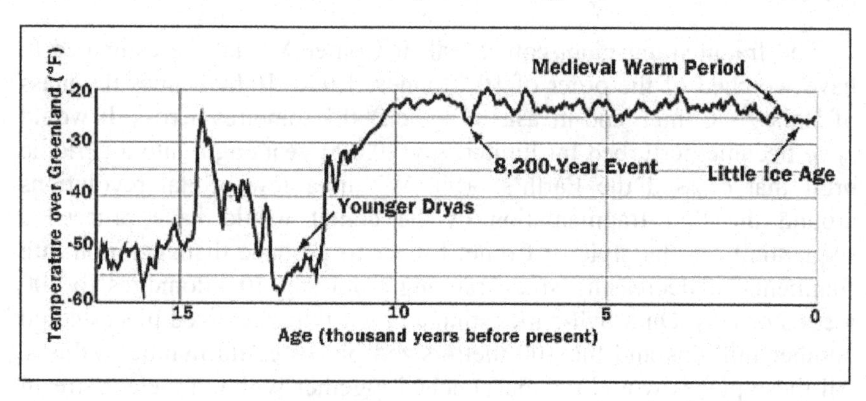

Fig. 13.1. Record of the Earth's average temperature over the past 17,000 years (Alley, 2002).

From Fig. 13.1 we note that the first stage of this process may have been caused by a collision with a large cometary fragment nearly 15,000 years ago. The Earth suddenly warms up but then cools down again to remain intermittently glaciated for the next 3,000 years, during which minor temperature oscillations may have been caused by a succession of much smaller impacts that dispersed vast quantities of dust into the stratosphere that reflected sunlight and prevented a stable green house from being established. Bill Napier has recently argued that some 12,900 years ago the Earth was indeed struck by thousands of 100-metre size cometary fragments over a relatively short timescale, leading to a dramatic cooling of the planet (Napier, 2010). The cooling, by as much as 8°C, interrupted the warming that had begun and caused glaciers to re-advance. It was exactly at this time, 12,900 years ago, a prehistoric Paleo-Indian group known as the Clovis culture suddenly disappears, and ice-age animals such as ground sloths, camels and mammoths also became extinct in North America. Recently a team of researchers led by James Kennett has discovered melt-glass material in a thin layer of sedimentary rock in Pennsylvania, South Carolina, and Syria, pointing to the effect of cometary missiles striking the Earth at this time.

The first significant and sustained temperature rise occurred at about 9500 BC, due perhaps to the collision of a slightly larger fragment of the comet. But even this one collision did not suffice to lift the Earth into a relatively stable interglacial epoch. There was then a sharp plunge back to ice age conditions and several sharp oscillations of temperature followed thereafter, most probably also caused by impacts.

A decisive emergence of the Earth from the last ice age had to await a major collision event 10,000 years ago. Water that was released due to evaporation from the oceans by the heat of impact was sufficient to restore the greenhouse effect on a very short timescale, thus causing the Earth to pass into a stable warmer phase.

From this moment in time, at about 8000 BC (10,000 years ago), the history of human civilisation might be said to have properly begun. At later times there is evidence of comparatively minor fluctuations in the Earth's average surface temperature over timescales of centuries to millennia. Fluctuations above and below the present-day value occur in the general range of 3–6 degrees F (~1.5–3 degrees C). It is hard to find a purely Earth-based mechanism that could satisfactorily explain such a pattern. But again impacts of cometary fragments provide a possible answer, and at any rate their contributions cannot be ignored. Comets, frequently breaking up in the high atmosphere or in the near-Earth environment, would inevitably produce an increase in the dust loading of the stratosphere, thus enhancing the reflective power of the atmosphere to solar radiation. The result is that less sunlight gets through the atmosphere to heat the surface, more is reflected back into space, and so cooling takes place. Thus, it is entirely natural that dips in the average temperature profile on the Earth's surface (Fig. 13.1) would be associated with atmospheric dusting events associated with episodes of bunched comet impacts.

For a typical cometary bolide of radius 40 metres hitting the Earth head-on at a speed of 14 km/s the kinetic energy of impact is equivalent to about 2 MT (megatonne) of the explosive TNT, that is, the equivalent of about 100 bombs of the type that destroyed Hiroshima in 1945. Such an object, if made of ice, would explode at about 30 kilometres above the surface of the Earth without any notable effect on human life. An object of only three times this size, a quarter of a kilometre in diameter, would strike much closer to the Earth and cause localised destruction on a significant scale, perhaps on the order of a city. A larger cometary bolide with a diameter of a kilometre would have an energy of 100,000 Hiroshima bombs and would cause widespread damage, perhaps on the scale of an entire country. So the size of the incoming missile is the crucial factor that determines what the eventual outcome of a collision would be. The bigger the object, the greater the damage it causes.

An object measuring about 100 metres across entered the upper atmosphere of the Earth over Tunguska in Siberia in the early hours of 30 June 1908. A great fireball was seen to pass low over the town of Kirensk and came down over a remote part of the Siberian taiga. The object did not reach the ground but exploded in the atmosphere at a height of about eight kilometres. The brilliant fireball, said to have outshone the Sun, was seen as far as 1,000 kilometres away from its point of descent, and the sound of the explosion was heard at even greater distances. Well-documented accounts of what happened are scant. The nearest to an eye-witness report is to be found in the Russian newspaper Sibir in Irkutsk, Siberia of 2 July 1908:

"Early in the ninth hour of the morning of June 30, a very unusual phenomenon was observed here. In the village of Nizhne Karelinsk (200 versts north of Kirensk) in the north-west sky very high above the horizon, the peasants saw a body shining very brightly, indeed too bright for the naked eye, with a blue-white light. It moved vertically downward for about ten minutes. The body was in the shape of a cylindrical pipe. The sky was cloudless, except that low on the horizon, in the same direction as the luminous body, a small black cloud was seen. The weather was hot and dry, and, as the luminous body approached the ground (which was covered by forest), it seemed to be crushed to dust, and in its place a vast cloud of black smoke formed, and a loud explosion, not like thunder, but as if from the avalanche of large stones or from heavy gunfire was heard. All the buildings shook, and at the same time a forked tongue of flame burst upward through the cloud. All the inhabitants of the village ran into the street in terror. Old women wept, everyone thought that the end of the world was upon them…"

The immense blast wave that resulted from the explosion felled trees over a distance of some 40 or 50 kilometres, and the heat from the fireball charred tree trunks for distances of up to 15 kilometres from the centre of impact. This scene of devastation, first photographed in 1927, is shown in Fig.13.2. Estimates of the total energy of the impacting object range from 13 to 30 megatonnes of TNT, equivalent to the explosive power of 650–1,500 Hiroshima bombs.

Fig.13.2. Tunguska site photographed in 1927.

Collisions of the Tunguska type, and others on a much grander scale, must surely have occurred sporadically as well as recurrently throughout our history and prehistory for the past 10,000 years. If bunched collision episodes were separated by intervals of 300–3,000 years, long periods of relative quiet may be seen to be interspersed by tens or hundreds of years of fierce bombardment.

The odds of a person being killed in a Tunguska-type event is easily calculated by dividing the lethal area around a point of impact, say an area of 5,000 square kilometres, by the total area of the Earth's surface, about 100 million square kilometres, giving an odds of one in 20,000.

If such impacts took place at a rate of once per year, the odds of perishing from a Tunguska-type impact would be roughly the same as it is today for death from an accident on our motorways, that is chance of about 10,000:1. But during an episode of bunched impacts with 100 impacts per year the chance over a life time of 30 years, say, would be about 30 percent. By far the most important consequence of such events is that over a bunched collision epoch, occupying say a century, during a bad spell, so to speak, more than one population centre in three would be totally destroyed. Every survivor would have observed what appeared like fire pouring down from the sky from a neighbouring conurbation, perhaps tens of kilometres away. It would be hard to imagine that experiences such as these would not have made an indelible impression on the minds of our ancestors and on their collective psyche. Their effect on the evolution of belief systems, religions and myths would have been profound.

At the dawn of civilisation some 10,000 years ago, cometary activity, including dust production, following the fragmentation of a progenitor comet, would have been conspicuously intense. For much of the year the entire zodiac may have glowed from the sunlight that was diffusely scattered by cometary dust particles. Sightings of comets ferociously breaking up, rising and setting, displaying their brilliant dust tails would thus have been exceedingly common in the ancient skies. There can be little doubt that myths and legends would have evolved in response to such experiences that must surely have been shared by many nomadic tribes that were scattered widely across our planet. With the power of the sky so manifestly clear it is easy to understand why the gods of most early societies were placed in the skies. One might also understand why such sky-centred religious systems often depicted celestial combat myths in which winged serpents and dragons were involved.

The rise and fall of civilisations, the ascendancy and decline of empires, that punctuate human history over the last 10,000 years, can be explained on the basis of periodic assaults from the skies. The falls of civilisations occur dramatically during short "bad periods" involving impacts, and the ascents to glory would have been maintained over longer periods of relative freedom from impacts. Fred Hoyle pointed out that a cometary connection might be inferred from the important discovery of metal smelting. This was indeed a remarkable discovery that led eventually to the use of metals for weapons, tools and equipment. It marked a crucial turning point in the economic fortunes of man. The idea of converting a piece of stone into malleable metal could surely not have occurred to anyone as an abstract concept. The discovery of metal smelting must therefore have been an unplanned accident, but then the challenge would be to understand how this same accident happened at the same time in widely separated locations on the Earth.

Archaeological evidence shows that copper was being used for making tools and utensils at a date close to 4300 BC. The first recorded use of pure copper is to be found in eastern Anatolia, but very quickly it spreads across the world. The indication is that the relevant natural accident that produced copper smelting had to be repeated in several locations almost simultaneously. This was almost certainly the multiple impacts of cometary bolides. Events of the Tunguska-type during episodes of bunched collisions could have started forest fares on a huge scale. Beneath the intense heat of glowing charcoal, rocks containing appropriate metallic ores would become naturally smelted. Our nomadic ancestors did not need any exceptional powers of observation to come across sites of smouldering forest fires similar to those that raged in Tunguska in 1908. They would only have had to pick up pieces of the smelted copper to discover that they could be beaten and flattened to yield artefacts that served their needs. This marked the beginning of the Copper Stone Age as it is sometimes called, which in turn led to the Bronze Age 1,000 years later.

Archaeological evidence for a 2800 BC event, although it exists, appears to be scant, but it is interesting to note that this time coincides with the end of the First Dynasty in Egypt. It is also the time when pyramid building commenced in a modest way with the construction of "step pyramids" that eventually evolved into the more impressive smooth-faced pyramid structures in later years.

A better documented episode or episodes of bunched collisions span the period 2500–2300 BC. Just before this time, several great civilisations and long-lived dynasties are known to have flourished, both in ancient Egypt and in the Indus Valley of North India. The ruined city of Mohenjo-daro in northern Pakistan was the site of a pre-Aryan civilisation that was perhaps more advanced than that of Egypt. It had flourished for over a millennium, but suddenly and dramatically collapsed. Aryan invasions from the West could have produced a slow erosion of an already aging empire, but not a seemingly cataclysmic fall. Likewise, seasonal flooding of the Indus Valley might have had a slow cumulative effect over many centuries but not a sudden one. A far more dramatic catastrophe could have arisen from tidal waves and tsunamis that follow quite naturally when cometary fragments crash into the sea. There has been recent evidence to suggest that a dust layer and a burnt surface horizon, apparently caused by an air blast, exist in archaeological sites of northern Syria dated at about 2350 BC.

Perhaps the strongest and most compelling evidence of an episode of cometary impacts is to be found in the deserts of Egypt. Following the unification of Upper and Lower Egypt by King Menes around 3100 BC, a desert empire flourished, reaching heights of glory, through a succession of dynasties of the so-called Old Kingdom of Egypt until its eventual collapse at about 2160 BC. Pyramid texts describe a prolonged period of turmoil preceding the collapse. The construction of the three most famous Giza pyramids began with Snefru's son Khufu who built the Great Pyramid around 2500 BC. This stupendous structure has a base that covers some 13 acres and a height of over 450 feet. The geometrical precision of the construction as well as the exact alignments of its faces to cardinal points (NSEW) is remarkable to say the least. Two other major pyramids were built at Giza over the next two centuries by Khufu's son Khafre and his successor Menkaure.

Why in the first place did the Egyptians choose to build such stupendous structures if they served no function except as royal tombs? Why did they scatter them over a vast area of desert, rather than build them all in one place? The Giza pyramids have survived for 4,500 years, surely withstanding several episodes of cometary bolide assaults from the skies. Fred Hoyle suggested that a solid pyramid structure may have just the right properties to maximize chances of surviving the blast wave from Tunguska-type explosions in the sky. One could not think of a better monument to survive millennia rather than decades or centuries.

Another theory is that the pyramids were not just royal tombs, nor were they simply religious icons, but they actually served the kings before their death, as air-raid shelters, to protect them from the explosions of Tunguska-type cosmic missiles. The channel within the Great Pyramid pointing in the direction of Orion's belt could, with the deployment of a single plane mirror or polished surface, have been used to observe the progress of an offending meteor stream, focusing on the point on the sky from which the missiles would have appeared to come. Other channels through the pyramid could have been used for other purposes: life support perhaps, ventilation, and transport of food. Archaeological discoveries by Donald B. Redford of Penn State University point to the remains of what may have been cosmic air-raid shelters for lesser mortals near the end of the Old Kingdom. Groups of human skeletons have been discovered with arms placed over their heads and bodies in various contorted positions, exactly as they fell, strongly suggesting that they were the hapless victims of an unexpected assault from the skies. The bodies are found to be associated with remnants of curved walls that look very much like parts of an air-raid shelter that did not survive.

More compelling evidence for a cosmic cataclysm that ended the Old Kingdom comes from work of scientists at Queen's University Belfast led by Mike Baillie. They used the new science of dendrochronology, which involves the study of the thicknesses of annual tree rings at different times in the past. (Each year a new ring is formed in the trunks of trees, resulting from growth during the summer months.) A narrower tree ring corresponding to a particular year means that there was little or no tree growth during that year, which could only be explained as arising from greatly diminished levels of sunlight. Such a thinning of tree rings in Irish oaks has actually been discovered over the entire period 2354–2345 BC, which comes close to the final decades of the Old Kingdom. This is easily explained as being due to the arrival of Tunguska-type cometary missiles that dusted the atmosphere and dimmed the light from the Sun.

The next well-documented episode of bunched collisions may have occurred a millennium later during the period 1350–1100 BC. Here again Baillie's evidence from tree ring thickness measurements shows a marked climatic downturn during 1159–1141BC, which could have been caused either directly by cometary missile impacts or by volcanoes that were triggered by such impacts.

The dates of events described in the Old Testament are open to dispute, but some at least may have occurred between 1300 and 1100 BC. Many of the Old Testament accounts of seemingly bizarre and mysterious occurrences could have had a basis in fact if one admits the possibility of an epoch of bunched cometary impacts. Descriptions of deluge, a rain of fire on the cities of Sodom and Gomorrah, famines occasioned by the wrath of the gods — all have a rational basis as possible effects of cometary impacts. Fires, tsunamis or tidal waves, floods, climatic changes adverse to crops, even clusters of earthquakes, can be interpreted as real phenomena caused by the arrival of cometary missiles. No metaphysical or mystical explanations are needed. We can also begin to understand what it was that Joshua saw when he reported that the Sun stood still in the sky. It was almost certainly the glow of an immense fireball, similar to what was seen over Tunguska in June 1908. The two sets of descriptions, in the Old Testament and in Siberia of 1908, are strikingly similar.

We have already noted that the break-up of comets during bunched collision episodes would have frequently led to spectacular displays in the ancient skies, and that these in turn would have given birth to celestial combat myths in ancient societies. Such celestial combat myths involving wars of the gods are clearly found in Greek traditions as in the Homeric poems around the eighth century BC. It is likely that Greek mythology evolved from the older mythologies of Western Asia and Mesopotamia that date back somewhat before 1100 BC when the Earth had indeed suffered an episode of bombardment by cometary bolides.

A mainly benign period appears to have started a few centuries before the classical period in Greece and continued with a few remissions until the dawn of the sixth century AD. Events in the sixth century AD bear all the hallmarks of an episode of cometary missile impacts, although perhaps less intense than at earlier times. The collapse of the Roman Empire in the sixth century AD has been the subject of intense scholarly debate for many years. Edward Gibbon's account of what happened leaves little room to doubt the reality of geological disturbances that played an important part in the final stages of the disruption of the Empire. The description goes thus:

"... history will distinguish ... periods in which calamitous events have been rare or frequent and will observe that this fever of the Earth raged with uncommon violence during the reign of Justinian (AD527–565). Each year is marked by the repetition of earthquakes, of such duration that Constantinople has been shaken above forty days; of such an extent that the shock has been communicated to the whole surface of the globe, or at least of the Roman Empire. An impulse of vibratory motion was felt, enormous chasms were opened up, huge and heavy bodies were discharged into the air, the sea alternately advanced and retreated beyond its ordinary bounds, and a mountain was torn from Libanus and cast into the waves ... Two hundred and fifty thousand people are said to have perished ... at Antioch."(*Decline and Fall of the Roman Empire*, Edward Gibbon)

The type of prolonged and frequent earthquake activity described by Gibbon is unusual. An explanation has to be sought in terms of an external cause — cosmic missile impacts that could provide the trigger to send pressure waves into the Earth's crust and thereby generate exceptionally prolonged bursts of seismic activity.

Evidence of another major downturn in the Earth's climate at precisely this time has again come from work by Baillie. His study of tree ring thicknesses in Irish oaks shows that there was little tree growth during the years around 540 AD. Similar studies by others have also shown the same effect — narrow tree rings at this time — in places as wide afield as Germany, Scandinavia, Siberia, North America and China. The idea that a volcanic eruption was responsible for a dust shroud that lowered the temperature and reduced seasonal tree growth does not tally with the lack of an acid signal in Greenland ice drills of the same period. Furthermore, volcanic dust is known to settle in a couple of years at most, and cannot therefore explain such a protracted episode (535–546 AD). So there can be little doubt that a major global catastrophe enveloped the planet around the year 540 AD.

From the beginning of the 7th century AD to the present epoch it would appear that there has been a largely benign period free of the traumatic events of cometary impacts. However, meteor events have been recorded in literary allusions as well as murals in the Dark Ages and the Middle Ages throughout much of Europe. The effect of variable comet and meteor dusting over the planet is also indicated by records of climate patterns. A medieval warm spell in Europe, during which grapes grew in southern England was followed by a series of cold spells between 1550 and 1850 AD, all pointing to variable dusting of the atmosphere due to the break-up of cometary bolides.

Chapter 14

The Mystery of the Red Rain

Descriptions of the rain of blood falling from the sky have been recorded in diverse cultures from very ancient times. Early literary allusions are to be found in Homer's Iliad, where Zeus twice caused blood to rain from the sky, and on one occasion did so to warn of imminent slaughter in battle. In Book 16 of Iliad it is recorded that in the midst of an episode of meteoric activity there was a shower of bloody rain. "Zeus noticing that his son Sarpendon would die sent a shower of bloody raindrops to the Earth in tribute." (McCafferty, 2008). The Greek historian Plutarch (47–120 AD) refers to rain of blood during the reign of Romulus, founder of Rome. Mike Baillie has cited an interesting remark attributed to Roger of Wendover written in 541 AD referring to a "comet in Gaul so vast that the whole sky seemed on fire. In the same year there dropped real blood from the clouds…" Similar ideas persisted through the Middle Ages and into the 17th and 18th centuries.

In classical Greece, events such as a shower of blood were interpreted as a demonstration of divine power, whilst in Christendom in medieval Europe people were less inclined to attribute such phenomena to supernatural causes, and natural explanations were often sought. Indian mythology also records similar events and likewise considers them to be omens and portents of the end of the world. In the classic epic Mahabharata the following account is given:

"The air was filled with the shouting of men, the roaring of elephants, the blasts of trumpets, and the beating of drums: the rattling of chariots was like to thunder rolling in heaven. The Gods and Gandharvas assembled in the clouds and saw the hosts which had gathered for mutual slaughter. As both armies waited for sunrise, a tempest arose and the dawn was darkened by dust clouds, so that men could scarce behold one another. Evil were the omens. Blood dropped like rain out of heaven, while jackals howled impatiently, and kites and vultures screamed hungrily for human flesh. The earth shook, peals of thunder were heard, although there were no clouds, and angry lightning rent the horrid gloom; flaming thunderbolts struck the rising sun and broke in fragments with loud noise."

McCafferty (2008) in a scholarly review of many historical sources collates many noteworthy allusions to the conjunction of meteoritic events and red rain-type descriptions. Although caution has to be exercised in assessing ancient documents the sheer weight of evidence is impressive and cannot be ignored in the author's view. Two impressive records cited by McCafferty are worthy of note.

The strongest link between red rain and a meteor fall is probably this example, from Egypt, 30 BC:

"Not only did rain fall in places where no drop had ever been seen before, but blood besides, and the flash of weapons appeared from the clouds, as the showers of blood mingled with water poured down. In other places the clash of drums and cymbals and the notes of flutes and trumpets were heard, and a serpent of enormous size suddenly appeared and uttered a hiss of incredible volume. Meanwhile comets were observed in the heavens..." (Dio, Book 51, xvii).

Chinese mythology also recounts a similar correspondence:

"The three Miao tribes were in great disorder and Heaven decreed their destruction. The sun came out at night and for three days it rained blood. A dragon appeared in the ancestral temple and dogs howled in the market place. Ice formed in summertime, the earth split open until springs gushed forth, the cereal crops grew differently, and the people were filled with a great terror..."

From the 19th century onward, there was a trend towards examining these events more scientifically. Christian Gottfried Ehrenberg (1795–1876), a German naturalist and professor of medicine at Berlin University conducted experiments at the Berlin Academy to recreate "blood rain" with a mixture of red dust and water. To the extent that many of the red rain incidents in the last 100 years may have been caused by red dust possibly from the Sahara, Ehrenberg may have been on the right track. But many red rain episodes in modern times had other causes, of which red biological cells — perhaps algae — were the most common. It is possible that red rain type-events are relatively common even in the present day but they go unnoticed and unrecorded except on occasions when the intensity of redness is particularly strong.

Popular as well as scientific interest in this phenomenon was revived in a dramatic way following events that took place on the morning of 25 July 2001 over a large area around the state of Kerala in India. A sonic boom heard over the area was followed by a fall of red rain — a story that is strikingly similar to those described in many historical accounts. The first red rain fall lasted for about 20 minutes, and this was repeated through the day, and also intermittently and episodically for nearly 8 weeks. Godfrey Louis, a physicist (currently of Kochin University) who collected and examined samples of the rain, was quickly able to dispose of the possibility of red dust being the cause of the redness. Figure 14.1 shows a light microscope image of a field of the red rain cells, which are clearly distinguishable from what dust particles would have appeared with their irregular shapes and a much broader spread of sizes. The presence of a cell wall seen even at low magnification and a translucent redness indicates clearly that it is a red pigment within a putative living cell. The average diameter of a red rain cell is about 5 micrometres.

From estimates of the total rainfall (measured in centimetres), the mass fraction of red cells in a typical red rain sample, and the area of incidence, the total mass of red cell material was estimated to be 50,000 kg (50 tonnes) (Louis and Kumar, 2003). Assuming further that the red rain material is 1 percent of the mass of a porous cometary bolide in which this material was dispersed, we can calculate its radius to be about 10 metres. If such a meteoroid possessed a loose fragile structure it can easily disintegrate in the high atmosphere releasing the red rain cells that eventually seed rain clouds in the troposphere. In this connection it is interesting to note that McCafferty (2008) cites a documented red rain event in October 1846 in France where the total mass of over 300 tonnes was recorded. In this instance about $1/8^{th}$ of the mass of the dispersed material was estimated to be in the form of microscopically identified diatoms.

With the strong influence of an Earth-centred viewpoint in biology it is not surprising to find any suggestion that the red rain cells may be extraterrestrial to be vigorously challenged. Sampath *et al.* (2001)

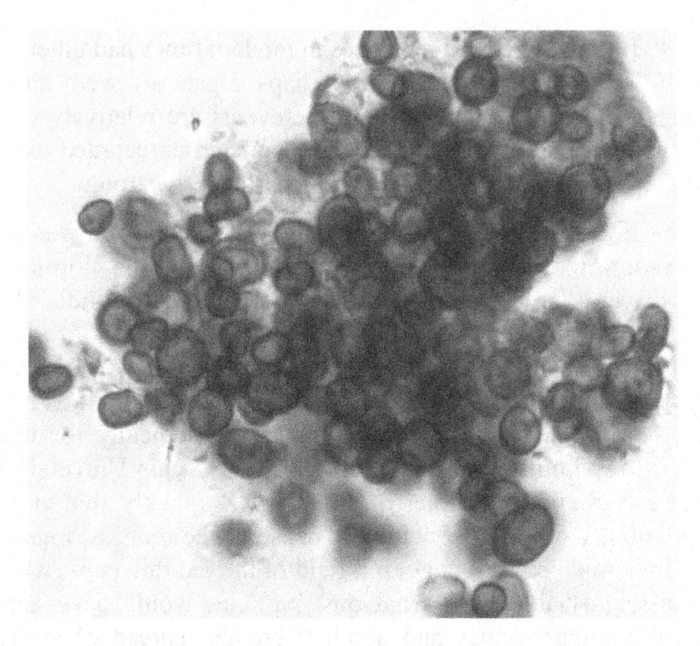

Fig. 14.1. Kerala red rain cells under optical microscope. Cell walls are visible as outer skins.

asserted with very little evidence that the Kerala red rain cells were none other than algae belonging to the genus *Trentepohlia*, a full 50 tonnes of this being lofted to the clouds from trees on Earth. This claim has not been substantiated in more detailed studies as we shall see later, but it has been widely cited as a refutation of a cosmic connection.

To counter such facile claims Louis and Kumar showed that the cells in the red rain are morphologically different from both ordinary algae and red blood cells, and this result has been independently verified. It has also been demonstrated that the red rain cells have a range of extraordinary properties which all point in the direction of an extraterrestrial origin.

Based on all the available laboratory data together with the observation that the first red rain event was preceded by a sonic boom, presumably caused by the explosion of a meteoroid, an extraterrestrial origin seems highly probable. The geographic pattern of incidence and the time distribution of the red rain cases do not fit readily with any terrestrial origin hypothesis, and appear to be more consistent with an origin from fragile cometary fragments that disintegrated in the upper atmosphere. The prolonged period of settling of small particles to be expected following an initial deposition in the upper atmosphere can explain the protracted episode of red rain events over several weeks. Alternatively it is possible that the terrestrial clouds provided a local habitat in which an initial injection of red rain cells on 25 June came to be episodically amplified in several discrete bursts.

In 2009 Godfrey Louis supplied the author with samples of the Kerala red rain and these were investigated by 3 PhD students Kani Rauf, Nori Miyake and Rajkumar Gangappa in various ways. The results of their studies formed a major component of three separate doctoral theses. Rauf's and Miyake's dissertations are lodged with Cardiff University, and that of Gangappa is with Glamorgan University. The situation that the red rain cells have defied identification after the work that led to 3 dissertations speaks for itself. If it was indeed a well-known algal cell such as *Trentepohlia* it would have been long since discovered.

The extensive investigations of the Kerala red rain carried out by Kani Rauf led to results generally similar to those obtained by Louis and Kumar (2003, 2007). Figure 14.2 shows a transmission electron microscope image. All cells show an exceptionally thick cell wall outlined by two darkly stained membranes, one internal and one external. The cell wall has an average thickness of some 6,000 Å. Many cells also have additionally a 2,000–3,000 Å thick protective exterior coat.

If the red rain cells have an astronomical origin we would expect them to have spectral properties that were common to interstellar material. Rauf obtained an infrared spectrum of red rain cells using a Fourier Transform Infrared Spectrometer (FTIR) shown in Fig. 14.3. We see here a series of absorption peaks centred at 2.9, 3.4, 6.0, 6.2, 6.8, 7.2, 7.7, 9.6, 10.3, 11.0, 12.4, 13.3, 18.5 and 21.0 μm, most of which interestingly match those of the so-called unidentified infrared bands (UIB's) in protoplanetary nebulae as shown in Table 14.1. Since we saw in an earlier chapter that these are regions where stars and planets form, the connection would appear to be significant. The ultraviolet spectrum of the red rain material also displays an absorption peak near 2,175 Å, which is characteristic of interstellar dust in galaxies, and indeed throughout a large fraction of the observable universe. This feature is indicated in the spectrum shown in Fig. 14.4.

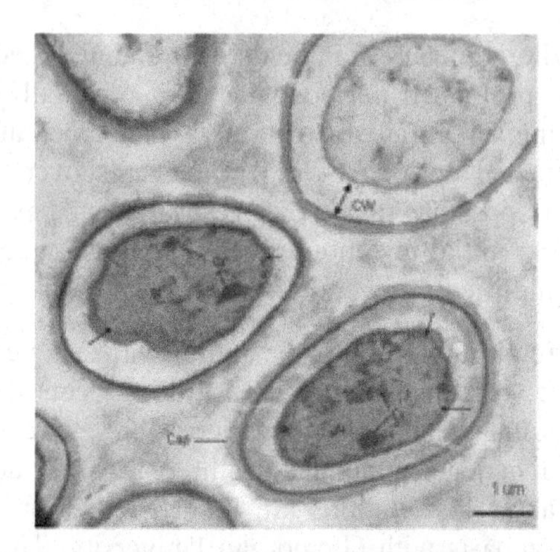

Fig. 14.2. Transmission electron microscopy of cross-sections of red rain cells (Rauf, 2012).

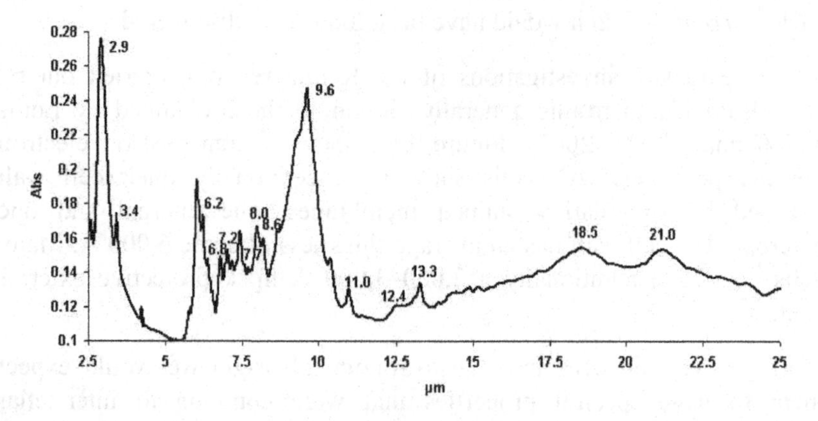

Fig. 14.3. FTIR spectrum of KBr-embedded red rain sample and distribution of infrared absorption peaks of red rain cells and astronomical emission bands.

Table 14.1. IR absorption peaks in µm: Protoplanetary nebulae compared with red rain.

PPN	3.3	3.4	6.2	6.9	7.2	7.7	8.0	8.6	11.3	12.2
RR cells		3.4	6.2	6.8	7.2	7.7	8.0	8.6	11.0	12.4

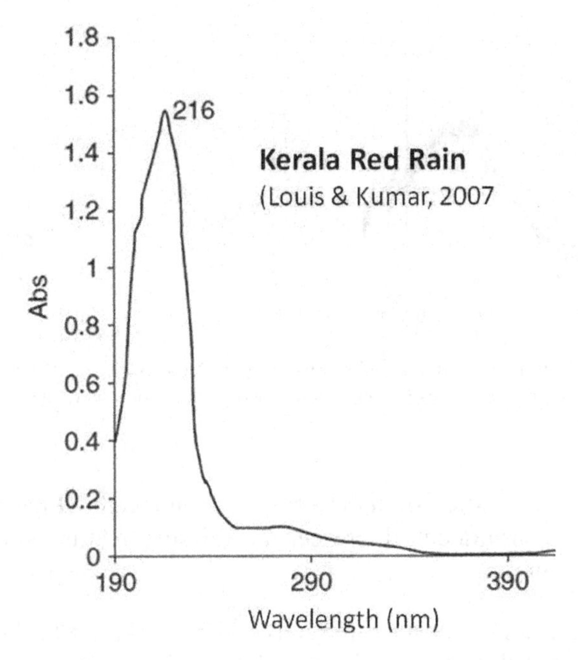

Fig. 14.4. Ultraviolet to visual spectrum of a sample of red rain showing a prominent peak centred at 2,070 Å.

We noted in Chapter 6 another property of interstellar dust that appears to be linked to a suite of complex biochemicals. This is the so-called extended red emission (ERE), a fluorescence behaviour associated with interstellar dust showing up in certain regions of the galaxy. This fluorescence property is observed in a wide range of astronomical objects where dust lies in close proximity to a source of exciting blue or ultraviolet radiation. We argued in Chapter 6 that the carriers of ERE are likely to be the same carbonaceous compounds as those responsible for the interstellar extinction feature at 2,175 Å. A similar connection may be sought in the red rain molecules producing the spectra of Fig. 14.5.

Figure 14.5 (left) shows the normalised extended red emission in NGC7023, a planetary nebula which is similar to many other sources in both the galaxy and the extragalactic sources. Although non-biological PAH explanations for this phenomenon are still being attempted, their success has so far been limited.

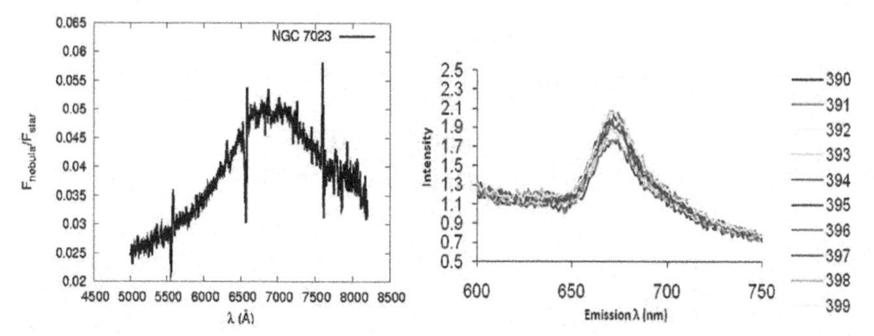

Fig. 14.5. Extended red emission from the NGC7023 (left) compared with fluorescence spectra of red rain cells with various excitation wavelengths (Gangappa *et al.*, 2010).

Fluorescence studies of the Kerala red rain cells (Fig. 14.5, right) have shown a prominent fluorescence emission feature similar to the ERE phenomenon.

Perhaps the single most contentious claim in relation to the properties of the Kerala red rain cells is that they replicate in a high pressure chamber at a temperature of 450°C (Louis and Kumar, 2006). At this temperature DNA is expected to break down, and other biopolymers may also be denatured. Although elemental analysis of the red rain cells showed 50% carbon, Louis and Kumar (2006) found no phosphorous in the cells. Phosphate groups form an important part of the DNA double helix, so the absence of phosphorous, if real, would imply that no DNA is present in the cells. This is what Godfrey Louis has maintained. Since DNA is the genetic material that carries the information content of life in all terrestrial cells, its absence could provide the strongest case for an extraterrestrial origin. One might speculate on the presence in the red rain cells of a non-DNA based genetic template of some kind. The challenge then would be to identify a DNA precursor that does not contain phosphorous or some other template that could hold and transmit the information content of the cells. Otherwise, replication and reproduction will be impossible.

Several investigators have found conflicting results with regard to the presence of DNA in red rain cells. Biologists use a stain called DAPI applied to cells to detect the presence of DNA. DNA in cell nuclei binds to the DAPI stain and so can be observed to fluoresce, by which effect a positive DNA content is inferred. These tests when applied to the red rain cells have on some occasions shown a positive result for DNA, but then the conflict with a "no phosphorus" result has to be resolved. The fact remains, however, that although several attempts have been made to isolate DNA from the red cells with a view to sequence it, so far they have all proved unsuccessful. The mystery thus deepens concerning the nature of these cells.

To add to the confusion, replication of the red rain cells has been independently verified, so the cells do indeed satisfy a basic requirement of biology — reproduction. A high temperature culture experiment has shown that the cells can be made to replicate in a standard nutrient medium at a temperature of 121°C and under high pressure conditions in an autoclave (pressure cooker) (Gangappa *et al.*, 2012). Even at room temperature, signs of a replication cycle, involving a budding process with daughter cells extruded from the cell wall, have been observed. This is shown in the transmission electron micrographs of Fig. 14.6.

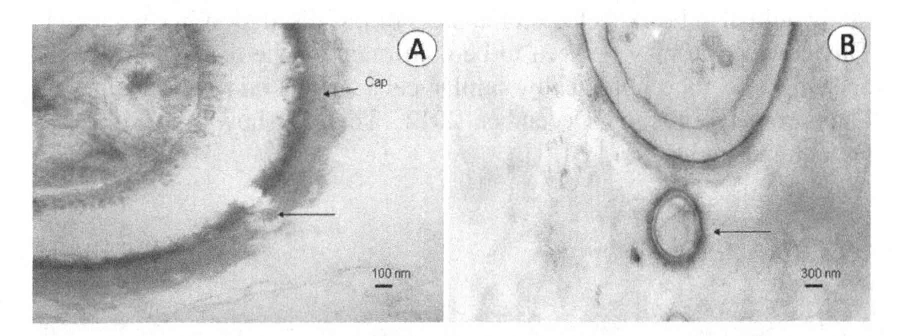

Fig. 14.6. Budding of daughter cells (Rauf, 2012).

By the end of 2012 our investigations on the red rain of Kerala had reached a dead end. A proposed identification with a relatively common algal species *Trentepohlia* has been shown to be wrong. Similarities are indeed evident at a superficial level of recognition, but differences are to be seen clearly in the fine structure. More significantly the DNA in *Trentepohlia auriga* is easily extracted and sequenced, whereas this has been impossible for red rain cells. All the indications are that the Kerala red rain cells represent an unknown microorganism of extraterrestrial origin.

Sri Lankan Red Rain

On the morning of 14 November 2012 the skies darkened over the ancient city of Polonnaruwa in Sri Lanka, and in the surrounding district red rain fell intermittently for several hours. The red rain of Kerala we discussed in the last section finally showed up in Sri Lanka. At earlier dates in November as well as in December there were numerous reports of fireball sightings in the area. Meteor activity generally tends to peak during this period with the Taurids, Geminids and Leonids being amongst the prominent regular showers. Such meteor showers result when the Earth in its orbit crosses the orbits of debris released from a particular comet, Comet Encke in this case for the Taurids.

David Ascher predicted that the Taurids would peak in intensity during this time in 2012 with the dominance of larger fragments, in a 61-year cycle. It is not surprising therefore that fireball sightings were more frequent during the Taurid shower of November 2012 and December 2013. The connection between meteors, freezing nuclei and rain would also be expected on the basis of our discussion in Chapter 11. This connection may be seen to be confirmed by the discovery of what appear to be morphologically similar cells to red rain cells in the Sri Lankan meteorites of December 2012. This is shown in Fig. 14.7 (Wickramasinghe *et al.*, 2013).

Fig. 14.7. Comparison of scanning electron micrographs of a field of Sri Lankan red rain cells with a field of similar structures in the Polonnaruwa meteorite.

On this basis it may be surmised that red rain cells dispersed by meteoroids may have seeded the red rain events in Sri Lanka.

Just as in Kerala there was a succession of similar, but less heavy red rain episodes that were distributed spatially over a few thousand square kilometres. The heaviest rains however took place in a smaller area and were centred around the original site of its incidence on November 14. Through the good offices of the Director of the Medical Research Institute in Sri Lanka, Anil Samaranayake, the author secured samples of the red rain material for study in the UK. These were analysed in Cardiff by Nori Miyake, who found the red cells to be very similar to the Kerala red rain cells. Elemental analysis confirmed the claim made earlier for the Kerala red rain that there was no phosphorous in the cells. Miyake also found that the outer layers of the cell wall contained uranium. Uranium is a rare element on Earth, so to find a terrestrial organism that can concentrate it and be lofted in vast quantities into the stratosphere appears unlikely. There were also independent studies done at the Sri Lanka Institute of Nanotechnology that confirmed the lack of phosphorous.

As with the Kerala red rain the jury is still out as to the origin of these cells but there are tantalising clues that point to their alien origin. If phosphorous is really absent in these cells there is a good chance that the biology carried with the red rain cannot interact with DNA-based living cells. It may be an alien life form that fell from the sky but did not take root on our planet.

Epilogue

The history of science has been punctuated by a succession of rebellions against authority. Whenever a conformist position becomes firmly established in science its eradication or even revision becomes exceedingly difficult irrespective of the weight of evidence. Sociology rather than science tends to prevail. The confrontations that arose throughout history were particularly fierce and protracted whenever they concerned matters that related to cosmology. In relatively recent times, in the 1960's, bitter disputes arose between supporters of the so-called standard Big Bang cosmology and the steady-state cosmology of Hoyle, Bondi and Gold. The dispute was ultimately resolved at least partially in favour of a broad class of Big Bang cosmologies principally due to the discovery of the cosmic microwave background. There is now little doubt that most of the material we observe in stars and galaxies had their origin in some form of Big Bang-type or explosive creation event, but within this general class of cosmology there remain several contending models. The idea of a quasi-steady-state universe was developed by Hoyle, Burbidge and Narlikar in the last decade of the 20th century (Hoyle *et al.*, 2000). Although this QSSC model is still not popular, it is one of several non-conformist models that could be reconciled with the available body of astronomical evidence.

The transfer of microbial life across galactic and extragalactic distances requires first and foremost the pre-existence of C, N, O, P and other elements needed for life in adequate quantity. This in turn implies access to regions of the Universe where star formation is under way. Such a condition is satisfied more or less uniformly in the spiral arms of our galaxy, and probably also in the galactic halo which has recently been shown to provide a feedstock of material for the formation of stars, and perhaps even for the storage and dispersal of life. Spectroscopic studies of external galaxies have also shown evidence of the presence of vast quantities of carbonaceous material suitable as a feedstock for life, including evidence of complex organic molecules in both the solid and the gas phase.

We have stressed in this book that a defining property of life is the superastronomical quantity of information that is held in the arrangements of its constituent molecules — amino acids and nucleotides in particular: amino acids in enzyme chains, and nucleobases in DNA and RNA. This crucial information for life, once it has been generated somewhere in a cosmological context, will be held within bacteria and viruses in a form that can be continually regenerated and re-assembled on countless planetary abodes. The most efficient intergalactic transport of biological information can be achieved with microbes attached onto iron whiskers of typical diameters of 0.02 μm and lengths about a millimetre, such as would condense naturally in expanding envelopes of supernovae, as metallic vapours cool (Hoyle and Wickramasinghe, 1988). These iron whiskers, along with their microbial hitch-hikers, are very strongly repelled by the infrared radiation from parent galaxies, reaching typical speeds of $\sim 10^4$ km/s in intergalactic space.

Even with the minutest fraction of surviving biological "messages" attached to fast moving whiskers in this way, biology could diffuse through a radius of ~50 Mpc of intergalactic space, a volume occupied by $\sim 10^6$ galaxies, in a mere Earth age, ~4.5 billion years (4.5 Gyr). Even partially corrupted genetic messages delivered via the agency of viruses or bacteria could serve as transmitters of the information of biology (Wesson, 2010).

The mistaken remit of modern astrobiology is to seek new origins of life — even against incredible odds — everywhere in cosmic locations thought to be life-friendly. This is not merely a trivialisation of what is, but a travesty of all the facts that are available. Wherever life originated in the universe and by whatever process, repeated multiple origins must be reckoned to be impossible and, moreover, irrelevant. The continuation and propagation of life are assured by the processes of panspermia as we have discussed in this book. With the most recent estimates of the total number of habitable planets running into hundreds of billions in our galaxy, the operation of panspermia is a foregone conclusion.

Although the so-called standard cold dark matter–dark energy cosmology is widely regarded as the final solution to the problem of the origin of the Universe, this confidence in the author's view is certainly premature. At every stage in the history of cosmology the belief was that the solution at hand was final and irrevocable. At every stage this confidence was shown to be in error, and so it would be prudent to keep an open mind on such big issues in science. Several alternative cosmologies consistent with the available observational data still remain in the field.

In the quasi-steady-state cosmology (Hoyle *et al.*, 2000) the universe expands exponentially with a time scale $P \approx 1000$ billion years while undergoing oscillations of shorter duration of $Q \approx 40$–50 billion years. New matter is created at the beginning of each oscillatory cycle at the epoch of maximum density, in such quantities as to maintain the same overall maximum density at the beginning of each cycle. This matter is processed during the cycle and undergoes physical as well as biological evolution. The essential point is that evolutionary processes proceed at a fast enough rate to transform this new matter into the required form in the cycle period of $Q \approx 50$ Gyr, and through ~20 such cycles, maintain a steady level over the expansion period P. To be consistent with modern astronomical observations the last such cycle must have just happened to begin 13.8 billion years ago.

If we assume that the biologically permeated region grows at such a rate as to double its size in 5 billion years (5 Gyr), then over a period of $Q \approx 50$ Gyr, the doubling process would go to a factor of $2^{10} \cong 10^3$, and over a period of $P \approx 1000$ Gyr it would grow by a factor of $\sim 10^{60}$. It is thus entirely possible for the quasi-steady-state universe to maintain a distribution of living systems at a steady level over the timescale ~P. We conclude that such cosmological models offer the best scope for a re-origination of life in the event of its annihilation, and for its maintenance throughout time. Whilst not being at variance with the range of astronomical data available at the present time, this remains a minority position in cosmology.

In this book we have placed emphasis on another minority viewpoint in cosmology, the hydro gravitational dynamics (HGD) model of Carl H. Gibson and Rudolf. E. Schild (2003). This model is consistent with the general class of QSS cosmology where the start of the last cycle is placed at 13.8 Gyr ago. The defining aspect of hydro gravitational dynamics (HGD) cosmology is that viscous forces cannot be neglected in the formation of cosmic structures, and that viscosity triggers fragmentation first at 30,000 years into 10^{49} g mass scales (thousands of galaxies) and later at 300,000 years into ~10^{46} g mass scales (galaxies). These protogalaxies comprised initially of plasma (ionized hydrogen) then fragment into two scales of mass when the plasma recombines to form a neutral gas. The material of protogalaxies some 13.8 billion–300,000 years ago comprised of two mass scales one at 10^{39} g and the other at 10^{27} g. The latter mass units form the primordial planets comprised mainly of H and He and enriched with the heavy elements that were synthesized when a small fraction coalesced to form massive stars and supernovae.

It is the high pressure, high temperature interiors of heavy element-enriched primordial planets that provide optimal conditions for life's origin (or re-origination within the QSS class). Given the large cosmological volumes available and the numerous panspermic interconnections between some 10^{80} primordial planets, these conditions will not have been repeated in later epochs. The time of such an origin is placed at between 2 and 8 million years after a Big Bang-type origin of the universe.

We have seen that astronomical data at infrared and ultraviolet wavelengths provide evidence of life-related molecules in galaxies of redshifts as high as $z = 5$, perhaps going back to a few hundred million years after the Big Bang. If the existence of near-solar relative abundances of carbon and other products of stellar nucleosynthesis can be interpreted as a prospector of life, such evidence is not lacking even in objects of higher redshift — closer to the Big Bang. One could infer from this and similar data that the products of stellar nucleosynthesis suited to serve as the feedstock of life existed in abundance all the way back to the highest observed redshifts. The cosmic horizons of life have thus been pushed embarrassingly close to the putative Big Bang origin in standard cosmologies, even before the first galaxies could have formed.

The ingress of alien microbial life onto our planet, whether dead or alive should not by any rational argument be perceived as a cause for concern. This is particularly so if, as we have discussed, a similar process of microbial injection has continued throughout geological time. Unlike the prospect of discovering alien intelligence, which might be justifiably viewed with apprehension, the humblest of microbial life forms occurring extraterrestrially would not constitute any threat. Neither would the discovery of alien microbes impinge on any issues of national sovereignty or defence, nor challenge our cherished position as the dominant life form in our corner of the Universe.

Over the past three decades we have witnessed a rapid growth of evidence for extraterrestrial microbial life. Along with this has grown a tendency on the part of scientific authorities to deny or denounce the data or even to denigrate the advocates of alien life. My own personal involvement in this matter dates back to the 1970's when, together with the late Fred Hoyle, I was investigating the nature of interstellar dust. At this time evidence for organic molecules in interstellar clouds was accumulating at a rapid pace, and the interstellar dust grains that were until then believed to be comprised of inorganic ices were shown by us to contain complex organic polymers of possible biological provenance (Wickramasinghe, 1974; Wickramasinghe *et al.*, 1977; Hoyle and Wickramasinghe, 1977a, b). These discoveries came as a surprise to astronomers, and for a long time the conclusion was resisted that such molecules might have a relevance to life on the Earth (Hoyle and Wickramasinghe, 1986).

Biologists in the 1960's and 1970's had no inkling of the intimate connection of their subject with astronomy. The holy grail of biology was the hypothesis that life emerged from a primordial soup generated *in situ* from inorganic molecules on the primitive Earth (Oparin, 1953). The hypothesis of the terrestrial origin of the chemical building blocks of life might have been thought plausible and necessary before it was discovered that vast quantities of biogenic organic molecules existed within the interstellar clouds (Hoyle *et al.*, 1978; Kwok, 2009). Having first argued for a complex biochemical composition of interstellar dust, Hoyle and the author were among the first to make a connection between complex organic molecules in interstellar clouds and life on Earth (Hoyle and Wickramasinghe, 1976, 1978, 1981). The total amount of organic material in the galaxy in the form of organic dust and PAH-type molecules account for about a third of all the carbon present in interstellar space — a truly vast quantity amounting to some billion or so solar masses (see review by Kwok, 2009).

The author's first inkling of any censorship relating to extraterrestrial life came when we made the intellectual leap from prebiology in space to fully-fledged biology outside the Earth (Hoyle and Wickramasinghe, 1976, 1982; Hoyle *et al.*, 1984). In setting out to explore the hypothesis that interstellar grains were not just abiotic organic polymers but bacterial cells in various stages of degradation, we made a prediction that interstellar dust in the infrared spectral region must have the signature of bacteria (Hoyle *et al*., 1982). This prediction was verified in a dramatic way when we discovered an amazingly close match to our predicted absorption curve for bacteria in the first infrared spectrum of an infrared source (GC-IRS7) near the galactic centre — a prediction that was made a full 3 months ahead of the serendipitous observations being made (Hoyle *et al*., 1982).

In Chapter 5 we have indicated how astronomical observations of interstellar dust and molecules ranging in wavelength from the far ultraviolet to the infrared have continued to support a biological origin. It would appear that a large fraction of interstellar dust grains and molecules must have a biological provenance, implying that microbial life exists on a grand galactic or even cosmological scale. In Chapters 12 and 14 the new evidence from the red rain of Kerala and Sri Lanka and the remarkable meteorite find in Sri Lanka provide stunning evidence in support of the cosmic theory of life.

The handicap facing 21st century science is an excess of specialisation. The polymaths of the nineteenth and twentieth centuries are a rare breed. Although some extent of specialisation is predicated by the huge quantity of information within each separate discipline, the disadvantage is that cross-disciplinary developments are discouraged. This, I believe, is one factor at least that has impeded the acceptance of the cosmic theories of life. To astronomers the association of bacteria with interstellar grains would appear understandably strange; to biologists the intrusion of astronomy into their discipline would be equally repugnant. But the Universe of course encompasses. In the year 2014 all the relevant facts appear to converge in favour of the cosmic origins of life. Resisting the facts and imposing censorship would in the long run be futile. The Universe will always have the last say.

Bibliography

A'Hearn, M.F. *et al.*, 2005. *Science*, 310, 258.

Abel, D.L. and Trevors, J.T., 2006. *Physics of Life Reviews*, 3, 211.

Abel, D.L., 2009. *Theor. Biol. Med. Model*, 6(1), 27. Open access at http://www.tbiomed.com/content/6/1/27.

Allen, D.A. and Wickramasinghe, D.T., 1981. *Nature*, 294, 239.

Alley, R.B., 2002. *The Two-Mile Time Machine: Ice Cores, Abrupt Climate Change, and our Future*, Princeton U. Press, Princeton.

Al-Mufti, S. *et al.*, 1983. In *Fundamental Studies and the Future of Science* (ed. C. Wickramasinghe), University College Cardiff Press, p. 342.

Alvarez, L.W. *et al.*, 1980. *Science*, 208, 1095.

Andrewes, C., 1965. *The Common Cold*, W.W. Norton, New York.

Arrhenius, S., 1903. *Die Umschau*, 7, 481.

Arrhenius, S., 1908. *Worlds in the Making*, Harper, London.

Asher, D.J. and Clube, S.V.M., 1993. *Quarterly J. Roy. Astro. Soc.*, 34, 481–511.

Baillie, M.G.L., 1994. *The Holocene*, 4, 212–217.

Baillie, M.G.L., 1996. *Acta Archaeologica*, 67, 291–298.

Bianciardi, G. *et al.*, 2012. *IJASS,* 13(1), 14.

Bidle, K. *et al.*, 2007. *Proc. Natl. Acad. Sci. USA*, 104(33), 13455.

Bigg, E.K., 1983. In *Fundamental Studies and the Future of Science* (ed. C. Wickramasinghe), University College Cardiff Press, p. 38.

Bohler, C. *et al.*, 1995. *Nature*, 376, 578.

Borucki, W.J. *et al.*, 2010. *Science*, 327, 977.

Boto, L., 2009. *Proc. R. Soc. B* 2010 277, 819–827.

Bowen, E.G., 1956. *Nature*, 117, 1121.

Brownlee, D.E. *et al.*, 1977. *Proc. Lunar Sci. Conf. 8th*, pp. 149–160.

Burchell, M.J. *et al.*, 2004. *Mon. Not. Roy. Astr. Soc.*, 352(4), 1273–1278.

Cairns-Smith, A.G., 1966. *J. Theor. Biol.*, 10, 53.

Cano, R.J. and Borucki, M., 1995. *Science*, 268, 1060.

Cassan, A. *et al.*, 2013. *Nature*, 481, 167.

Cataldo, F., Keheyan, Y. and Heymann, D., 2002. *Int. J. Astrobiol.*,1, 79.

Claus, G. and Nagy, B., 1961. *Nature* 192, 594.

Claus, G., Nagy, B. and Europa, D.L., 1963. *Ann. NY Acad. Sci.*, 108, 580.

Clemett, S.J. *et al.*, 1993. *Science*, 262, 721.

Clube, S.V.M. *et al.*, 1996. *Astrophysics and Space Science*, 245, 43–60.

Clube, V. and Napier, W.M., 1990. *The Cosmic Winter*, Basil Blackwell, Oxford.

Cockell, C.S., 1999. *Planetary and Space Science*, 47, 1487.

Cole, A.E., Court, A. and Kantor, A.J., 1965. Model atmospheres, in *Handbook of Geophysics and Space Environments* (ed. S.L. Valley), Air Force Cambridge Research Laboratories, p. 22.

Crick, F.H.C. and Orgel, L.E., 1973. *Icarus*, 19, 341.

Crovisier, J. *et al.*, 1997. *Science*, 275, 1904.

Darbon, S., Perrin, J.-M. and Sivan, J.-P., 1998. *Astron. & Astrophys.*, 333, 264.

Darwin, C. and Wallace, A.R., 1858. *Zool. J. of the Linnean Soc.*, 3, 46.

De Groot, N.G. *et al.*, 2002. *Proc. Natl. Acad. Sci. USA*, 99, 11748–11753.

Deamer, D., 2011. *First Life*, University of California Press, Berkeley, California.

Draine, B.T., 2003. *Ann. Rev. Asron. Astrophys.*, 41, 241.

Dressing, C.D. and Charbonneau, D., 2013. *Astrophys. J.*, 767, 95.

Elíasdóttir, Á. *et al.*, 2009. *Astrophys. J.*, 697, 1725–1740.

Franck, S. *et al.*, 2003. *Int. J. Astrobiol.*, 2, 35.

Furton, D.G. and Witt, A.N., 1992. *Astrophys. J.*, 386, 587.

Gangappa, R., Wickramasinghe, C., Wainwright, M. *et al.*, 2010. *Proc. SPIE*, 7819, 78190N1.

Gangappa, R., 2012. PhD Thesis, University of Glamorgan, UK.

Gibson, C.H. and Schild, R.E., 2009. *Appl. Fluid Mech.*, 2(2), 35–41. arXiv:0808.3228.

Gibson, C.H., Schild, R.E. and Wickramasinghe, N.C., 2011. *Int. J. Astrobiol.*, 10(2), 83–98.

Gibson, D.G. *et al.*, 2010. *Science*, 329, 52–56.

Gould, S. and Eldridge, N., 1977. *Paleobiology*, 3(2), 115–151.

Gregory, R.H. and Monteith, J.L. (eds.), 1967. *Airborne Microbes — Symposium for the Society of General Microbiology*, Vol. 17, Cambridge University Press.

Haldane, J.B.S., 1928. *Possible Worlds*, Hugh and Bros, New York.

Haldane, J.B.S., 1954. *New Biology*, 16, 12.

Harris, M.J., Wickramasinghe, N.C., Lloyd, D. *et al.*, 2002. *Proc. SPIE*, 4495,192.

Hoover, R.B., 2005. In *Perspectives in Astrobiology* (eds. R.B. Hoover, A.Y. Rozanov and R.R. Paepe), IOS Press, Amsterdam, pp. 43–65.

Hoover, R.B., 2011. *Journal of Cosmology*, 13, 3811–3848.

Horie, M., Honda, T., Suzuki, Y. *et al.*, 2010. *Nature*, 463, 84–87.

Horneck, G., Mileikowsky, C., Melosh, H.J. *et al.*, 2002. In *Astrobiology. The quest for the conditions of life* (eds. G. Horneck, C. Baumstark-Khan), Springer, Berlin.

Hoyle, F., 1993. *Origin of the Universe and the Origin of Religion*, Moyer Bell, Rhode Island.

Hoyle, F., Burbidge, G. and Narlikar, J.V., 2000. *Alternative Cosmology*, Cambridge University Press.

Hoyle, F. and Wickramasinghe, N.C., 1962. *Mon. Not. Roy. Astr. Soc.*, 124, 417.

Hoyle, F. and Wickramasinghe, N.C., 1969. *Nature*, 155, L181.

Hoyle, F. and Wickramasinghe, N.C., 1976. *Nature*, 264, 45.

Hoyle, F. and Wickramasinghe, N.C., 1978. *Lifecloud: The origin of life in the Universe*, J.M. Dent, London.

Hoyle, F. and Wickramasinghe, N.C., 1979. *Diseases from Space*, J.M. Dent, London.

Hoyle, F. and Wickramasinghe, N.C., 1981. *Evolution from Space*, J.M. Dent, London.

Hoyle, F. and Wickramasinghe, N.C., 1982. *Proofs that Life is Cosmic*, Memoirs No. 1, Institute of Fundamental Studies, Sri Lanka. www.panspermia.org/proofslifeiscosmic.pdf.

Hoyle, F. and Wickramasinghe, N.C., 1986a. *Nature*, 322, 509.

Hoyle, F. and Wickramasinghe, N.C., 1986b. *Earth, Moon and Planets*, 36, 289.

Hoyle, F. and Wickramasinghe, N.C., 1990. *J. Roy. Soc. Med.*, 83, 258–261.

Hoyle, F. and Wickramasinghe, N.C., 1991. *The Theory of Cosmic Grains*, Kluwer Academic Press, Dordrecht.

Hoyle, F. and Wickramasinghe, N.C., 2000. *Astronomical Origins of Life: Steps towards Panspermia*, Kluwer Academic Press, Dordrecht.

Hoyle, F., Wickramasinghe, N.C. and Al-Mufti, S., 1984. *Astrophys. Sp. Sci.*, 98, 343.

Hoyle, F., Wickramasinghe, N.C. and Pflug, H.D., 1985. *Astrophys. Sp.Sci.*, 113, 209.

Jain, R., Rivera, M.C., Moore, J.E. *et al.*, 2003. *Mol. Biol. Evol.*, 20(10), 1598–1602.

Johnson, F.M., 1971. *Ann. New York Acad. Sci.*, 194, 3.

Johnson, F.M., 1972. *Ann. NY Acad. Sci.*, 187, 186.

Jones, B.W. *et al.*, 2005. *Astropys. J.*, 622, 1091.

Joseph, R. and Wickramasinghe, N.C., 2011. *Journal of Cosmology*, 16, 6832–6861.

Kajander, E. and Ciftcioglu, N., 1998. *Proc. Natl. Acad. Sci. USA*, 95(14), 8274.

Kasten, F., 1968. *J. App. Meteorology*, 7, 944.

Keeling, P.J. and Palmer, J.D., 2008. *Nature Reviews Genetics*, 9, 605–618.

Kopparapu, R. *et al.*, 2013, *Astropys. J. Lett.*, 767(1), L8.

Kristensen, L.E., van Dishoeck, E.F., Tafalla, M. *et al.*, 2011. *Astron. & Astrophys.*, 531, L1. arXiv: 1105.4884v1.

Kwok, S., 2009. *Astrophys. Sp. Sci.*, 319, 5–21.

Lage, C.A.S. *et al.*, 2012. *Int. J. Astrobiol.*, 11(4), 251.

Levin, G. V. and Straat, P. A., 1976. *Science*, 194(4271), 1322–1329.

Lindahl, T., 1993. *Nature*, 362, 709.

Lisse, C.M., van Cleve, J., Adams, A.C. *et al.*, 2006. *Science*, 313, 635.

Louis, G. and Kumar, A.S., 2006. *Astrophys. Sp. Sci.,* 302, 175.

Manning, C.E., Mojzsis, S.J. and Harrison, T.M., 2006. *Am. J. Sci.*, 306, 303.

Matsuoka, K., Nagao, T., Mailino, R. *et al.*, 2011. *Astron. & Astrophys.*, 532, L10.

Mattila, K., 1979. *Astron. & Astrophys.*, 78, 253.

Mayor, M. and Queloz, D., 1995. *Nature*, 378, 355.

McCafferty, P., 2008. *Int. J. Astrobiol.*, 7(1), 9–15.

McKay, D.S. *et al.*, 1996. *Science*, 273, 924.

Mileikowsky, C., Cucinotta, F.A., Wilson, J.W. *et al.*, 2000. *Icarus*, 145, 391.

Miller, S.L., 1953. *Science*, 117, 528.

Miller, S.L. and Urey, H.C., 1959. *Science,* 130, 245.

Miyake, N., Wallis, M.K. and Al-Mufti, S., 2010. *Journal of Cosmology*, 7, 1743.

Mojzsis, S.J. *et al.*, 1996. *Nature*, 384, 55–59.

Mojzsis, S.J., Harrison, T.M. and Pidgeon, R.T., 2001. *Nature*, 409, 178.

Morowitz, H. and Sagan, C., 1967. *Nature*, 215, 1259.

Morrison, P. and Cocconi, G., 1959. *Nature*, 184, 841.

Motta, V., Mediavilla, E., Muñoz, J.A. *et al.*, 2002. *Astrophys. J.*, 574, 719–725.

Nagy, B. *et al.*, 1963. *Nature*, 193, 1129.

Nandy, K., 1964. *Publ. Roy. Obs. Edin.*, 4, 57; 3,142.

Nandy, K., Morgan, D.H. and Houziaux, L., 1984. *Mon. Not. Roy. Astr. Soc.*, 211, 895.

Napier, W.M., 2010. *Mon. Not. Roy. Astr. Soc.*, 405, 1901–1906.

Napier, W.M., Wickramasinghe, J.T. and Wickramasinghe, N.C., 2007. *Int. J. Astrobiology*, 6(4), 321–323.

Narlikar, J.V., Wickramasinghe, N.C., Wainwright, M. *et al.*, 2003. *Current Science*, 85(1), 29.

Noterdaeme, P., Ledoux, C., Srianand. R. *et al.*, 2009. *Astron. & Astrophys.*, 503, 765–770.

Ohno, S., 1970. *Evolution by Gene Duplication*, Allen & Unwin, London.

Oparin, A.I., 1938. *The Origin of Life*, Macmillan, London (Original Russian book 1924).

Orgel, L.E. and Crick, F.H.C. 1968. *J. Mol. Biol*, 7, 238.

Pasteur, L., 1857. *C.R. Acad. Sci.*, 45, 913–916.

Perrin, J.-M., Darbon, S. and Sivan, J.-P., 1995. *Astron. & Astrophys.*, 304, L21.

Pflug, H.D. and Heinz, B., 1997. *Proc. SPIE*, 3111, 86.

Pflug, H.D., 1984. In *Fundamental Studies and the Future of Science* (ed. N.C. Wickramasinghe), University College Cardiff Press.

Ponnamperuma, C. and Mark, R., 1965. *Science*, 148, 1221.

Rauf, K. and Wickramasinghe, C., 2010. *Int. J. Astrobiol.*, 9(1), 29–34.

Russell, C.T. and Vaisberg, O., 1983. In *Venus* (ed. D.M. Hunten *et al.*), Univ. Ariz. Press, pp. 873–940.

Russell, C.T. *et al.*, 1982. In *Comets* (ed. L.L. Wilkening), Univ. Ariz. Press, p. 561.

Ryan, F.P., 2004. *J. R. Soc. Med.*, 97, 560–565.

Sagan, C. and Khare, B.N., 1971. *Science*, 173, 417.

Sampath, S. *et al.*, 2001. Colored Rain: A report on the phenomenon, CESS-PR-114-2001, Centre for Earth Science Studies, Thiruvananthapuram.

Sattler, B. *et al.*, 2012. *Geophys. Res. Lett.*, 28(2), 239.

Schild, R.E., 1996. *Astrophys. J.*, 464, 125.

Schmitt-Kopplina, P. *et al.*, 2010. *Proc. Natl. Acad. Sci. USA*, 107(7), 2763.

Schopf, J.W., 2006. *Phil Trans. R. Soc. B*, 361, 869.

Schopf, J.W., 1999. *Cradle of Life: The discovery of Earth's earliest fossils*, Princeton University Press.

Schulze-Makuch, D.H. and Irwin, L.N., 2002. *Astrobiology*, 2, 197.

Schulze-Makuch, D.H. *et al.*, 2004. *Astrobiology*, 4, 11.

Schwartz, R.N. and Townes, C.H., 1961. *Nature*, 190, 205.

Sharov, A.A., 2010. *Journal of Cosmology*, 5, 833–842.

shCherbak, V.I. and Makukov, M.A., 2013. *Icarus*, 224(1), 228.

Shivaji, S., Chaturvedi, P., Begum, Z. *et al.*, 2009. *Int. J. Systematic and Evolutionary Microbiology*, 59, 2977–2986.

Sivan, J.-P. and Perrin, J.-M., 1993. *Astrophys. J.*, 404, 258.

Smith, J.D.T., Draine, B.T., Dalie, D.A. *et al.*, 2007. *Astrophys. J.*, 656, 770.

Sumi, T. *et al.*, 2011. *Nature*, 473, 349.

Szomouru, A. and Guhathakurta, P., 1998. *Astrophys. J.*, 494, L93.

Tepletz, H.I., Desai, V., Armuo, L. *et al.*, 2007. *Astrophys. J.*, 659, 941–949.

Trevors, J.T., Pollack, G.H., Saier, Jr., M.H. and Masson, L., 2012. *Theor. Biosci.*, 131(2), 117–23. doi: 10.1007/s12064-012-0154-3.

Van de Hulst, H.C., 1949. *Recherche Astron. Utrecht*, 11, 2.

Van de Kamp, P., 1962. *Vistas in Astronomy*, 26(2), 141.

Vanysek, V. and Wickramasinghe, N.C., 1975. *Astrophys. Sp. Sci.*, 33, L19.

Venter, J.C.J., Adams, M.D., Myers, E.W. *et al.*, 2001. *Science*, 291, 1304–1351.

Vladimir, C. and Makukov, M., 2013. *Icarus*, 224(1), 228–242.

Vreeland, R.H., Rosenzweig, W.D. and Powers, D., 2000. *Nature*, 407, 897.

Wachtershauser, G., 1990. *Proc. Natl. Acad. Sci. USA*, 87(1), 200.

Wainwright, M., Wickramasinghe, N.C., Narlikar, J.V. *et al.*, 2003. *FEMS Microbiol. Lett.*, 218, 161.

Wallis, M.K., Wickramasinghe N.C., 2004. *Mon. Not. Roy. Astr. Soc.*, 348, 52–57.

Wallis, J. *et al.*, 2013. *Proc. SPIE*, 8865, 886508-1.

Wallis, J., Miyake, N., Hoover, R.B. *et al.*, 2013. The Polonnaruwa meteorite: oxygen isotope, crystalline and biological composition, *Journal of Cosmology*, 22, 10004–10011.

Wang, X., Mitra, N., Secundino, I. *et al.*, 2012. *Proc. Natl. Acad. Sci. USA*, doi: 10.1073/pnas.1119459109.

Wesson, P., 2010. *Sp. Sci. Rev.*, 156(1–4), 239–252.

Wickramarathne, K. and Wickramasinghe, N.C., 2013. *Journal of Cosmology*, 22, 10075–10079.

Wickramasinghe, C., 2010. The astrobiological case for our cosmic ancestry, *Int. J. Astrobiol.*, 9(2), 119–129.

Wickramasinghe, C., 2011. Bacterial morphologies supporting cometary panspermia: a reappraisal, *Int. J. Astrobiol.*, 10(1), 25–30.

Wickramasinghe, C., 2011. Viva Panspermia! *The Observatory*, 131, 130.

Wickramasinghe, D.T. and Allen, D.A., 1980. *Nature*, 287, L93.

Wickramasinghe, D.T. and Allen, D.A., 1986. *Nature*, 323, 44.

Wickramasinghe, J.T., Wickramasinghe, N.C. and Napier, W.M., 2010. *Comets and the Origin of Life*, World Scientific, Singapore.

Wickramasinghe, N.C., 1967. *Interstellar Grains*, Chapman and Hall, London.

Wickramasinghe, N.C. and Wickramasinghe, J.T., 2008. *Astrophys. Sp. Sci.*, 317, 133.

Wickramasinghe, N.C., 2012. DNA sequencing and predictions of the cosmic theory of life, *Astrophys. Sp. Sci.*, doi: 10.1007/s10509-012-1227-y.arXiv: 1208.5035.

Wickramasinghe, N.C., Hoyle, F. and Lloyd, D., 1996. *Astrophys. Sp. Sci.*, 240, 161.

Wickramasinghe, N.C., Lloyd, D. and Wickramasinghe, J.T., 2002. *Proc. SPIE*, 4495, 255.

Wickramasinghe, N.C., Samaranayake, A., Wickramarathne, K., Wallis, D.H., Wallis, M.K., Miyake, N., Coulson, S.J., Hoover, R., Gibson, C.H. and Wallis, J.H., 2013d, Living diatoms in the Polonnaruwa meteorite — Possible link to red and yellow rain, *Journal of Cosmology*, 21, 40.

Wickramasinghe, N.C., Wallis, J., Miyake, N., Wallis, D.H., Samaranayake, A., Wickramarathne, K., Hoover, R. and Wallis, M.K., 2013c. Authenticity of the life-bearing Polonnaruwa meteorite, *Journal of Cosmology*, 21, 39.

Wickramasinghe, N.C., Wallis, J., Wallis, D.H., Samaranayake, A., 2013a. Fossil diatoms in a new carbonaceous meteorite, *Journal of Cosmology*, 21, 37.

Wickramasinghe, N.C., Wallis, J., Wallis, D.H., Schild, R.E., Gibson, C.H., 2012. Life-bearing primordial planets in the solar vicinity, *Astrophys. Sp. Sci.*, doi: 10.1007/s10509-012-1092-8.

Wickramasinghe, N.C., Wallis, J., Wallis, D.H., Wallis, M.K., Al-Mufti, S., Wickramasinghe, J.T., Samaranayake, A. and Wickramarathne, K., 2013b. On the cometary origin of the Polonnaruwa meteorite, *Journal of Cosmology*, 21, 38.

Willner, S.P., Russell, R.W., Pietter, R.C. *et al.*, 1979. *Astrophys. J.*, 229, L65.

Willner, S.P., Soifer, B.T., Russell, R.W. *et al.*, 1977. *Astrophys. J.*, 217, L121.

Witt, A.N. and Schild, R.E., 1988. *Astrophys. J.*, 325, 837.

Woese, C. and Fox, G., 1977. *Proc. Natl. Acad. Sci. USA*, 74(11), 5088.

Woese, C., 1967. *The Genetic Code*, Harper and Row, New York.

Wu, M. *et al.*, 2005. *PLoS Genet.* 1(5), e65.

Index

Printed in the United States
By Bookmasters